Indoor Air Quality

A Guide for Facility Managers

Second Edition

Indoor Air Quality

A Guide for Facility Managers

Second Edition

Ed Bas

THE FAIRMONT PRESS, INC.
Lilburn, Georgia

MARCEL DEKKER, INC.
New York and Basel

Library of Congress Cataloging-in-Publication Data

Bas, Ed, 1954-
    Indoor air quality : a guide for facility managers/Ed Bas.
        p. cm.
    Includes bibliographical references and index.
    ISBN 0-88173-327-X (print) -- ISBN 0-88173-465-9 (electronic)
    1. Heating. 2. Ventilation. 3. Air conditioning. 4. Indoor air pollution. 5. Air quality management. I. Title

    TH7015.B38 2004
    697.1--dc21

                                                    2003056514

Published by The Fairmont Press, Inc.
700 Indian Trail, Lilburn, GA 30047
tel: 770-925-9388; fax: 770-381-9865
http://www.fairmontpress.com

Learning Resources
Centre

I 2 S 8 8 0 b 7

Distributed by Marcel Dekker, Inc.
270 Madison Avenue, New York, NY 10016
tel: 212-696-9000; fax: 212-685-4540
http://www.dekker.com

Printed in the United States of America

10   9   8   7   6   5   4   3   2   1

0-88173-327-X (The Fairmont Press, Inc.)
0-8247-4009-2 (Marcel Dekker, Inc.)

Contents

Preface

IAQ: A LONG ROAD FOR THE PAST 7 YEARS

Indoor Air Quality: A Guide For Facility Managers is written for facility managers, mechanical engineers, maintenance personnel, indoor air quality consultants, and other professionals interested in identifying problems and finding solutions for common indoor air quality. Written in the language of the layperson, it is designed to be used as a management tool, training guide, and ready reference for existing and future managers.

It's been 7 years since this book was first published. The industry has changed, and indoor air quality has presented many challenges. In the meantime, my career has grown. I was associate editor with *The Air Conditioning, Heating and Refrigeration News*. I have served as editor and publisher with *Snips*, a 70-year-old trade magazine that holds a classical domain in the fields of forced warm air, air conditioning, and sheet metal, including duct work, metal roofing and ornamental work.

This second edition of *Indoor Air Quality: A Guide for Facility Managers* has been slow in its creation, but is a product of its era. Maintenance problems are not always easily solved. In the 7 years since this book's first edition, strides have been made, but many potential solutions have been stalled. Mold has been a major problem and a key maintenance issue for many facility managers and building owners. The annual IAQ Conference continued by the American Society of Heating, Refrigerating and Air Conditioning Engineers (ASHRAE) almost (but not quite) called it the Mold Conference. It's the subject of a public session at Chicago's McConnick Place in January 2003. Joseph Lstiburek said, "Mold has been around a long time; what is the big deal now?" and "Buildings don't dry as quickly as they used to." He cited two of 13 items in "top 10 rules": 1) Buildings should be suited to their environment, 2) The laws of physics must be followed.

Often buildings are not suited to their environment, and engineers try to make them work outside the laws of physics.

A successfully operated building has a facility manager with the training and knowledge of how a building operates. A good manager fulfills roles involving inspection, maintenance, and cleaning, making sure that the building's airways and pathways are unobstructed. He must be sympathetic and responsive to owners, employees, and tenants about IAQ concerns.

On a personal level, my children, Derek and Shannon, have been a great part of my inspiration and stamina, helping to build an extraordinary home life that surrounded my writing efforts. They are loved.

Oh, I have an interest which keeps me busy… a stroke and 10-day hospitalization introduced me to aphasia on Saturday, July 11, 2002. My book editors have been near at hand during this time. My neurologist and therapist say it will take some time and effort. I know that. From Tips for Life After Stroke: "Expect change, it's not always bad. Things may be different and different can be okay." And "Keep your sense of humor. Look for reasons to laugh, laughing helps lighten your load."

Ed Bas

Foreword to the Second Edition

INDOOR AIR QUALITY IN COMMERCIAL BUILDING MAINTENANCE AND OPERATION

Building owners owe it to their tenants to provide the safest, healthiest indoor environment possible. This is something that soon will be taken for granted, right along with adequate parking space, elevators that don't break down, lobbies that are attractive, and office spaces that are workable.

A survey by the Building Owners and Managers Association (BOMA) International asked building tenants to rate the importance of comfort, the quality of their indoor air, acoustics, quality of building maintenance, building management's responsiveness, etc. IAQ scored high, but what was surprising is just *how* high. In fact, tenants were unanimous (99 percent) in rating IAQ, acoustics, noise control, and comfortable temperatures as among the most important features considered in their evaluation of office space. Only a handful of other features in the survey rated as highly. Rental rates, operating expenses and security, which also rated highly, by contrast were named by 89 percent, 86 percent, and 89 percent of respondents, respectively.

Unfortunately, it isn't always clear how to obtain or maintain this important feature.

Just as unclear is how a building's maintenance staff and outside service contractors are considered in this picture, as well as the impact the building's occupants have on their own surroundings

The science and study of indoor air pollution is a relatively new one. Airborne dust and high humidity, as examples, are nothing new to humans; but, suddenly these all-too-common and generally considered rather benign contaminants are suddenly gaining wider public attention, right alongside their more urgent and serious counterparts such as lead, radon, and asbestos.

A recent web surf for the words indoor air quality on the Internet produced nearly 9,000 "hits" and that number is growing daily.

The federal Environmental Protection Agency (EPA) ranks poor indoor air quality as one of the top five environmental issues in the United States. In fact, according to the EPA, typical airborne pollutants run two to five times higher indoors than out. The average adult breathes in 13,000 liters of air per day. Children breathe 50 percent more air per pound of body weight than adults, which increases their susceptibility to respiratory problems. An unhealthy office, school or nursing home correspondingly contributes to an unhealthy life-style, since we spend 90 percent of our time indoors.

A study in the 1980s asked 350 northern New Jersey home owners to wear personal air monitors to find the quality of the air they breathed on a typical day. For 11 target chemical studies, including some known carcinogens, the indoor concentrations were found to be 10 to 100 times more concentrated indoors than out. Said EPA scientist Lance Wallace, "The tests showed that the most pressing threat from air pollution isn't from the petrochemical or rubber plant out at the edge of town—it's from the paints and adhesives and other toxic products that we bring into our own homes."

Healthy Buildings International conducted 813 building studies between 1980 and 1992 relating to indoor air quality, involving 750,000 building occupants. It found that three-quarters (75.5 percent) were due to operating faults and/or poor maintenance. Fifty-four percent were due to poor ventilation prompted by energy conservation methods; 20.8 percent were due to poor air distribution; 56.9 percent were due to inefficient filtration; and 12.2 percent were due to contamination inside the ductwork. Dust and allergenic fungi comprised more than half (60 percent) of the most significant pollutants found, with low relative humidity another top-rated problem at 18.5 percent; formaldehyde, 8.5 percent; and tobacco smoke, 2.8 percent.

It's not only good business to be able to market a healthier indoor environment, it may soon be the law. Building owners must become aware of the liabilities involved in ignoring threats to indoor air. To deliver a building's occupants free from harm, at least to the best of current operational practices, is the key. Evidence exists that better indoor air quality can also improve energy efficiency and thus lower operating costs, as well as extend equipment life and improve individual productivity.

Tenants and occupants of commercial buildings are bringing IAQ cases against a broad array of defendants associated with building construction or maintenance, from building owners and managers (including insurance companies that own real estate) to contractors, engineers, consultants, and leasing agents. An expanding tier of lawsuits will affect architects, builders, real estate agents, subcontractors, manufacturers, etc.

Claims so far have targeted HVAC systems that over- or under-heated buildings, failed to cool buildings adequately, or failed to adequately ventilate whole buildings, or portions of them. The four most common types of projects in these cases have been commercial and industrial buildings, schools and university buildings, condominiums and hospitals. There have been some attempts by the insurance industry to offer IAQ coverage. But to this point, IAQ as a risk management issue has not been adequately addressed by the mainstream insurance industry. Hard data on IAQ claims is lacking, and the insurance industry thrives on such data. Instead, damaging costs are mostly hidden in the form of increased health and medical benefits, and unemployment insurance. These tend to be treated as part of the costs of doing business rather than as costs that are preventable and circumventable.

IAQ has at least one highly visible poster child. One of the most famous (or infamous) incidents of sick building syndrome involved a new courthouse in Polk County, Florida. After completion of the building, occupants soon began complaining of extensive indoor air quality problems, resulting in personal ailments, and building maintenance problems. Renovation costs and other building-related damages eventually rose to $40 million—about what it cost to build the structure in the first place, and not including lost productivity and resulting medical costs. It ultimately involved almost a total rehab, including new furnishings, replacement of woodwork and marble, and a thorough duct cleaning.

A courthouse in Martin County, Florida, built in 1989, had a similar experience, and the resulting renovations cost $24 million.

Although businesses can buy specific types of insurance to handle asbestos and lead paint problems, most insurance companies now write their commercial general liability policies in a way that excludes many IAQ-related claims. Some insurance companies have written their policies to specifically limit their exposure in IAQ-related cases.

Meanwhile, lawyers are escalating the "legal arms race," looking for other precedents on which to base IAQ suits. The most common areas of law they use today are:

(1) Contracts and breach of lease,
(2) Professional malpractice or negligence,
(3) Strict liability, often used against product manufacturers,
(4) Fraud or misrepresentation,
(5) Punitive damages.

Cases based on the provisions of the Americans with Disabilities Act (ADA) are also on the rise, resulting in claims related to multiple chemical sensitivity (MCS) syndrome. While many insurers and medical experts do not recognize MCS as a disease, the U.S. Department of Housing and Urban Development determined in 1992 that MCS is a "handicap" under the Fair Housing Act. Although the legal status of MCS is still far from clear, some courts interpreting ADA law have found that MCS can be called a "disability" under ADA, and this may open the door to ADA-based IAQ claims. This could mean that anyone with a particular aversion to, for example, airborne dust or volatile organic chemicals would have to be accommodated.

There are other reasons for improving building IAQ. A safe, healthy environment can be marketed as such. Many employers would welcome such space for their workers.

The savvy building owner who is aware of IAQ's pitfalls and has taken proper steps towards cleaner air should underscore those steps in marketing campaigns and leasing contract language. It is a valid way to distinguish one building from another, outside and above the usual location, landscaping and price-cutting. Indoor air quality doesn't have to be a mystery, even if all the answers aren't in yet. Much of it relies on common sense housekeeping and equipment specifying-operations practices that are already well known—even if not always practiced.

The commercial office building is a unique product, a product that is intended to last for the ages, a product that is unique and not mass produced, a product that is near and dear to those who buy it, lease space in it, and dwell in it. To builders, excellence is an issue that distinguishes one builder from another. Attention to construction detail is equally apparent in providing quality indoor air, and in taking the proper steps to preserve it.

Sick building syndrome is a term that has gotten popular only within the last decade or so, and experts are still debating its cause and cures. Some researchers and medical practitioners believe that a building can indeed make you sick, from poor indoor air quality.

But exactly how do you control the quality of the air in a commercial office building, school, or institution? Should tenants be concerned that the building they work in or do business in can actually be a source of illnesses and health complaints? Is your child's asthma or allergies really caused by "something in the air" that could be avoided?

According to the Environmental Protection Agency, all of us face a variety of risks to our health as we go about our day-to-day lives. Driving in cars, flying in planes, engaging in recreational activities, and being exposed to environmental pollutants all pose varying degrees of risk. Some risks are simply unavoidable. Some we choose to accept because to do otherwise would restrict our ability to lead our lives the way we want. And some are risks we might decide to avoid if we had the opportunity to make informed choices. Indoor air pollution is one risk that we can do something about. Often, the "cure" is based on simple prescriptive specifications—operation and maintenance practices that really should be common knowledge.

While some of the research in this area remains inconclusive, according to the EPA, "In the last several years, a growing body of scientific evidence has indicated that the air within homes and other buildings can be more seriously polluted than the outdoor air in even the largest and most industrialized cities."

Because people spend so much of their time indoors, it makes sense to try to make our indoor environments as healthy and productive as possible.

Chapter 1

Sick Buildings

There is no such thing as a "sick building," but it is a useful term to describe buildings that are unsafe or unhealthy for those who occupy them. Facility managers need to start an indoor air quality plan to identify, correct, and prevent indoor air quality problems. In many cases, taking action can limit exposure to liability, increase productivity, and save energy.

INTRODUCTION

Indoor air quality (IAQ) is a relatively new issue in the workplace. In the past few years, media accounts of problems with our indoor air have alerted facility managers to a potential crisis. Many of these concerns over the air we breathe can be directly linked to the building's mechanical system. Often, when the heating, ventilation and air conditioning (HVAC) system of a building is malfunctioning or simply is not well maintained, indoor air problems result.

Not all of these potential problems are new ones. Past workplace surveys conducted by the Building Owners And Managers Association (BOMA) International, Washington, DC, almost unfailingly put HVAC at the top of management, operation and design problems. This is not surprising to those facility managers in the trenches well familiar with barrages of complaints such as "It's too hot in here," "It's stuffy" or "This room is drafty." In fact, workplace surveys usually list the five top stressors among office workers as lack of air movement, being too hot in the summer, stagnant air, cigarette smoke and being too cold in the winter.

It is in HVAC that we concentrate most of our effects to improve IAQ.

People want to be comfortable to work indoors. Today, indoor comfort is something we have come to take for granted. Comfort is a component of IAQ, which also takes into account the presence of contaminants and bacteria that can cause illness. High numbers of cases of illness or tip-offs such as increased absenteeism can classify an entire building as "sick." On a broader scale, ergonomics, lighting and acoustics all tend to play important roles in the indoor environment, which can affect occupant perception of air quality.

A Healthy, Pleasant Environment

Our goal in ensuring quality indoor air is to provide a healthy, pleasant indoor environment where people are comfortable and productive. Failing to properly address these problems, however—particularly when the environment may be the cause of health complaints—runs the risk of lawsuits, poor public image, poor tenant relations or lost tenants (if the building is a rental property), lost work time and lower productivity. The good news is that a healthy building is not one that is very difficult to attain, nor is it necessarily one that is more expensive to operate.

IAQ problems range from the simple to the complex, and there are as many solutions as there are causes. Many IAQ problems boil down to the HVAC system's design failing to meet current needs, or due to improper HVAC operation and maintenance.

DEFINING THE "SICK BUILDING"

A common, useful definition of "sick building syndrome" is when at least 20 percent of the people occupying a building experience symptoms of illness for a period of two weeks or longer, but the source of the symptoms cannot be determined. In this term, "syndrome" is used to describe a number of symptoms occurring together. In 1982, the World Health Organization (WHO) listed eight specific symptoms associated with sick building syndrome, identified below. Note that not all of these symptoms must be present to classify a building as "sick."

- Irritation of eyes, nose and throat.
- Dry mucous membranes and skin.

Figure 1-1. Reprinted from *The Air Conditioning, Heating And Refrigeration News*, **Business News Publishing Co. Courtesy: Dan Saad.**

- Erythema (redness due to inflammation).
- Mental fatigue, headaches.
- Airway infections, cough.
- Hoarseness of voice, wheezing.
- Unspecified hypersensitivity reactions.
- Nausea, dizziness.

WHO has estimated that as many as one-third of the world's buildings today can be considered "sick." According to the U.S. Environmental Protection Agency (EPA) and WHO, 60 percent of American buildings have serious (20 percent) or somewhat serious (40 percent) problems with IAQ.

DEFINING GOOD INDOOR AIR QUALITY

The American Society of Heating, Refrigerating and Air-Conditioning Engineers (ASHRAE), the 100+-year-old Atlanta-based professional society for the HVAC industry that is often referred to in this book, defines acceptable IAQ in its Standard 62-1989 as "air in which there are no known contaminants at harmful concentrations as determined by cognizant authorities and with which a substantial majority (90 percent or more) of the people exposed do not express dissatisfaction."

This should not be confused with, "If nobody complains, the air is acceptable." There is seldom immediate evidence of a smoking gun with an IAQ problem—there are no canaries to take down into the mine shafts with us.

We must also take into account, as stated in the ASHRAE definition, the many possible unnoticeable pollutants and contaminants that may not be causing immediate problems but can be unacceptable, or even deadly, in the long run. Certainly, when employees in an office building begin to complain, then call in sick because of real or perceived health-related complaints, the facility manager can expect a reaction, and both management and occupants in turn will be expected to respond quickly.

Any falloff in productivity can be costly, of course, but to an even larger extent employers will be extremely wary of the more tangible problem of liability.

Elements Of Good Indoor Air Quality

As early as 1916, ASHRAE had laid the groundwork for achieving good IAQ. In that year, J.J. Blackmore, in his ASHRAE paper, "Can We Standardize The Requirements for Ventilation?" wrote:

...the art of ventilation calls for several operations beyond that of conveying into a building a quantity of fresh air and the removal of a similar quantity of (so-called) foul air from the same building... The other operations necessary for good ventilation may be designated as follows: air cleansing, relative humidity, temperature, air movement and odor removal.

Not a bad basis for our modern IAQ objectives. Today, 80 years later, we are grappling with these same issues in different ways.

The U.S. EPA defines elements of good IAQ as:

1. Ensuring adequate ventilation (introduction and distribution of clean indoor air).

2. Controlling contaminants traveling in the air.

3. Maintaining acceptable thermal comfort.

Ventilation - The HVAC system must introduce and distribute fresh outside air in suitable quantities to maintain an adequate supply of fresh air and to dilute contaminants. The amount of air introduced, in cubic feet per minute (cfm), should be based on ASHRAE Standard 62 guidelines for a minimum ventilation rate of 15-20 cfm per person, a substantial increase from the rate of 5 cfm per person in the previous 1973 and 1981 standards.

Air filters should be properly selected and maintained to remove particles and contaminants from the air before it is introduced (or reintroduced/recirculated) and distributed indoors. Air distribution should be adequate so that clean indoor air reaches the occupants and to avoid drafts and temperature gradients.

We will attempt to further define and address these needs in this book, but measuring or determining what is adequate airflow or "quality" indoor air is not always easy; a good measure of this relies on common sense or the absence of general complaints that could be attributable to poor IAQ. It is not simply a matter of setting a thermostat at 70 degrees and troubleshooting the system only when the temperature drops below or exceeds that comfort level.

ASHRAE approves to IAQ standard

Four addenda to the American Society of Heating, Refrigerating and Air-Conditioning Engineers' ventilation standard have been approved. ANSI/ASHRAE Standard 62-2001, "Ventilation for Acceptable Indoor Air Quality," sets minimum ventilation rates and other requirements for commercial and institutional buildings. Addendum 62af changes the purpose and scope of the standard to describe how it relates to new and existing buildings, clarifies its coverage of industrial and laboratory spaces, and adds a caveat concerning situations where outdoor air quality may be poor. The addendum states that the standard is intended for regulatory application to new buildings, additions to existing buildings, and those changes to existing buildings that are identified in the body of the standard.

The rate of air exchange from outdoors to interior spaces should be balanced, a process called demand control. If more air is leaving the building than is entering it through the outside air damper, pressure drop can occur. Interior air pressure will become negative, drawing air from the outside through all available gaps and leakage points in the building envelope. This is undesirable because the outside air may itself be contaminated, and the air entering the building will not be filtered or humidified.

Controlling Contaminants - Today's buildings may have as many as 900 contaminants indoors with thousands of sources—including new furniture, cleaning agents, smoking, new building materials, pesticides and even perfume and other cosmetics. Many contaminants are microbiological or otherwise organic, triggering asthma and allergies. In each cubic foot of air, there are millions of pollutants—such as bacteria, pollen spores and dust mites.

While increased ventilation will help to dilute these contaminants, certain indoor areas that produce undesirable levels of contaminants or odors can be isolated to the outside without circulating to other interior spaces.

In addition, key HVAC components must be properly maintained on a scheduled basis to remove dirt, dust and debris that can breed bacteria.

Thermal Comfort - The HVAC system should provide an acceptable level of comfort in temperature according to ASHRAE Standard 55-1981 guidelines. Humidity, which is integral to perception of temperature, must also be factored, and should be controlled in a range based on ASHRAE Standard 55. If relative (local) humidity is too high, an otherwise suitable temperature will feel too hot and stuffy to many people; the converse is also true, requiring unnecessary additional heating. Humidity extremes can also produce environments that can cause high levels of mold, mildew and bacteria growth—and make people sick (see Chapter 6).

Arguments for a Holistic Approach

Poor ventilation is often blamed as the culprit and increased ventilation is often hailed as a panacea for curing IAQ ills. Both are only part of the big picture. ASHRAE Standard 62-1989 in-

Figure 1-2. Most IAQ problems either have chemical or microbiological contaminants at the source. Courtesy: CH2M Hill.

creased the minimum ventilation rate of 5 cfm first established in 1973 to 15 cfm per person, depending on the type of building based on its activity—a smoking lounge, for example, would need much more ventilation than a mostly vacant auditorium. As a result, most facilities constructed or renovated from 1973-1989, including the large population of office buildings built in the 1970s, are now considered inadequately ventilated. That does not mean an inadequately ventilated building is always prone to IAQ problems, however. Ventilation is important to dilute indoor contaminants with clean outside air, when those contaminants are present. But to some facilities, paying attention to those other factors mentioned above is as important as addressing ventilation.

But back to Standard 62. Why the increase? To judge a standard, we must consider the time frame around which it was established. Standard 62-2001 was released two years after an often-quoted study released by the National Institute of Occupational Safety And Health (NIOSH). That study, which covered 446 buildings with IAQ problems, revealed that "inadequate ventilation" was responsible for more than half (53 percent) of IAQ problems. Contamination from furnishings and building materials were also cited (unknown causes were attributed to 13 percent of the problems). With "inadequate ventilation" the cause of more than half of IAQ problems, it was only logical to increase the minimum ventilation rate and since then, many people confuse increased ventilation with a single cure-all for IAQ ills.

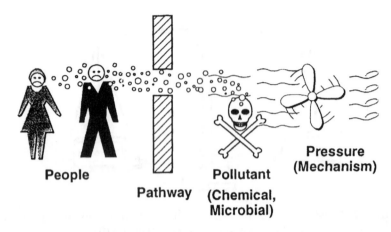

People   **Pollutant**   **Pressure (Mechanism)**

Pathway   **(Chemical, Microbial)**

Figure 1-3. Poor IAQ conditions becomes a people problem when there is a pollutant inside or outside the building that has a pathway to an occupied area. Often, the pollutant is delivered directly to people by the building's HVAC system. Courtesy: CH2M Hill.

Upon closer inspection, however, the study shows more than immediately meets the eye.

One out of five buildings (21 percent) was operating with no fresh air at all. Fifty-seven percent had inefficient filters—15 percent of these because of improper installation—and 44 percent of the buildings had dirty ventilation systems. Many of the ventilation problems, therefore, were not necessarily from lack of ventilation (except for the few that had no ventilation at all), but from lack of proper filtration and maintenance.

Similarly, Health Buildings International (HBI), Fairfax, VA, reported that out of 813 building studies it carried out between 1980 and 1992, involving 750,000 building occupants, three-fourths (75.5 percent) of IAQ problems were traced to operating faults and/or poor maintenance. Fifty-four percent were due to poor ventilation prompted by energy conservation, 20.8 percent due to poor air distribution, 56.9 percent due to inefficient filtration and 12.2 percent due to contamination inside the ductwork. Dust and allergenic fungi comprised more than half (60 percent) of the most significant pollutants or factors found, with low relative humidity 18.5 percent (high humidity was 2.1 percent), formaldehyde 8.5 percent, tobacco smoke 2.8 percent and ozone one percent.

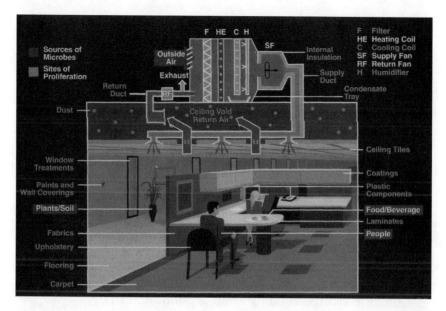

Figure 1-4. Microscopic organisms that can make people sick can originate at a variety of sources indoors, and proliferate in the immediate environment and the building's ventilation system.

It is clear from these studies that simply pumping in more outside air may not be the single, or even the best, cure for all cases. A flexible, holistic approach is often the best strategy for dealing with IAQ problems, incorporating all aspects of ventilation as well as thermal comfort. We must focus on operation and maintenance of the HVAC system as well as its design.

Additionally, although in this book we will deal often with handling the problem at the HVAC system, we must not overlook controlling contaminants as an equally important means of improving IAQ. The real problem quite often lies as its source, not in how it is handled.

BENEFITS AND POTENTIAL BENEFITS OF GOOD INDOOR AIR QUALITY

A relatively recent phenomenon, sick buildings are an ailment we can all do without—more troublesome than the flu, as frightening as the plague, as expensive as any civil law suit stemming from negligence that might be encountered.

The courts and government are sympathetic to employees who are forced to work in environments that can do them harm; this no longer applies only to industrial-type facilities, but also to hospitals, schools and office buildings. Employers and occupants react in kind to building owners and facilities managers who fail to do their part in committing to healthy buildings, and the reaction can be particularly severe when a lease agreement is involved.

If we take action to monitor IAQ and improve our indoor air, we can:

- Avoid penalties for non-compliance with safety regulations.

- Minimize exposure to liability and avoid litigation, which can be especially costly if the building's insurance policy does not include claims resulting from a polluting activity.

- Create a pleasant working environment that can maximize productivity and minimize absenteeism.

- Avoid negative publicity.

- If managing a rental property, avoid unpleasant relations with tenants and gain a marketing advantage to gain new tenants.

- Realize energy savings from improvements to the HVAC system.

Some of these benefits of good IAQ (and hazards of poor IAQ) merit more explanation, while some remain questionable as to their real impact on building performance and market value.

Comply with the Full Spirit of Safety Regulations

In 1980, requests to evaluate indoor office environments made up only eight percent of investigations made by NIOSH. But by 1992, that figure had grown to 75 percent.

Poor indoor air may qualify as an unseen health hazard under various laws, codes and regulations, including HR 1066 IAQ Act (1989, revised); AB 2249 Workplace Safety Act (1991); and SB 198 Injury And Illness Prevention Program (1991). The latter is considered the most encompassing, and violating it could result in a fine of $2,000-$7,500 for a first offense.

Minimize Exposure to Liability

Lawsuits may be the driving force behind many indoor air quality complaints and remediation efforts. Indoor air affects us all: We spend most of our time indoors, and much of this at work. Yet homeowners will often tend to be much more tolerant of their indoor air, especially if the likely remedies to unhealthy air or poor ventilation are costly to them. Increasing ventilation can cause energy costs to rise; energy recovery ventilators, better filtration and various air cleaners are available, but again, always at a cost. It is easier for most people to require this of a building owner, or one's employer, than to do it where it hits their own pocketbooks—that is, at the residential level.

Attorney Laurence S. Kirsch, esq., described it this way to the International Facility Management Association (IFMA) of Houston, TX in a conference paper:

> **After years of relative obscurity, the subject of indoor air pollution has now gripped the attention of the media and the public. The press and television often report on alleged "sick building" episodes in which various individuals in a single building experience symptoms such as coughing, headaches, respiratory irritation, dizziness or nausea that appear on entering the building and disappear on leaving.**
>
> **Building owners and operators are the ones to whom most complaints initially will be directed. They will be forced to evaluate whether these complaints warrant a response, and if so, what kind of response. In making these decisions, they will have to keep in mind that if building occupants do not receive what they believe to be a satisfactory response, they are likely to take the next step. In many cases, this will mean calling a lawyer.**

Anything a facility manager can do ahead of time regarding his indoor environment can either reduce the risk of a problem occurring, or at least may well be perceived as an act of due diligence in the courtroom. The stakes may be especially high if the building owner's insurance policy does not cover pollution-related claims.

"To me, due diligence is the process by which a building owner or manager provides tangible assurance of acceptable indoor air quality," said Jay Kirihara, vice president of QIC, an IAQ mitigation firm based in Dallas. "To a building owner or facility manager, indoor air quality takes on a special meaning. Facility managers must be able to respond to the question: 'What are you doing to assure acceptable indoor air quality?'"

Court cases at present are few and far between, but the number is growing. The main defendants are usually building owners and facility managers.

In one of the first IAQ court cases in the United States, Call vs. Prudential, a noteworthy ruling was that any individuals involved in the design or construction of a building could be held liable for IAQ-related injuries. The lawsuit was settled out of court in 1992 after five years of legal negotiations.

In another landmark decision at the end of 1993, Bahura vs. SEW Investors, the federal judge awarded five employees of the U.S. EPA $950,000 for health problems resulting from working in a facility in Washington, DC. The facility was a three-story main mall flanked by two 12-story office towers owned by SEW Investors. Compensatory damages ranging from $120,000 to $240,000 were awarded, while no punitive damages were awarded.

Jim Dinegar, vice president of government and industry affairs at the time for BOMA, proclaimed: "The floodgates have opened." He said there was still a great deal of confusion about what could be done to ensure safe IAQ: "We feel frustrated in not knowing what we will have to do to avoid future lawsuits." He said he considered the dollar amount substantial in a down market and at such an early stage for such legal decisions.

Fortunately for SEW Investors, a judge later rescinded all but one of the awards to the plaintiffs.

The Productivity Debate

Higher productivity resulting from improved IAQ is often used as an argument for incurring increased value that could, but not necessarily, come from improving indoor air. According to the U.S. EPA, IAQ-related health problems cost more than $1 billion

Figure 1-5. IAQ in the news. The issue has gripped public and industry attention in recent years. Articles reprinted from *The Air Conditioning, Heating and Refrigeration News*. Courtesy: Business News Publishing Co.

each year in medical costs and $60 billion each year in lower productivity and lost work time. The good news is ensuring good IAQ may improve productivity and bring some of this money back to the bottom line. Some researchers claim that they have measured improvements of 10 percent or more.

Consider the relationship between profit and productivity. If we estimate that an average salary for an office worker is $30,000 and that person occupies about 150 sq.ft. of office space, the salary cost related to space is $200/sq.ft. If productivity drops 10 percent due to environmental conditions, that means the employer can lose $20 per sq.ft. per year. Conversely, if productivity increases 10 percent as the result of improved environmental conditions, each worker can add $3,000 to the bottom line each year.

Harold Lorsch, Ph.D., PE and Ossama Abdou, Arch.D., in their ASHRAE paper titled, "The Impact of the Building Indoor Environment on Occupant Productivity," stated: "Since the cost of the people in an office is an order-of-magnitude higher than the cost of maintaining and operating the building, spending money on improving the work environment may be the most cost-effective way to improve worker productivity."

So far, however, research on workplace environments related to productivity is highly subjective and short-term. Measuring improvements and attributing those numbers to specific programs or criteria is difficult at best. Output of individual workers is known to fluctuate wildly, from 50-200 percent at any given time, due to causes not always attributable to the physical work environment. Some believe that external factors play much more of a role—conditions at home, for instance. An unhappy family life is likely to influence a worker's attitudes at work more than any fluctuation in, for example, humidity. Privacy, the relationship with the boss, amount of workload and nature of workload and other factors play understandably important roles. For this reason, some researchers believe that there is little reason to argue that ASHRAE should continue supporting additional research studying possible links between the indoor environment and worker productivity.

Based on some studies, however, other researchers are satisfied that a correlation exists.

One such study, conducted by Larry C. Holcomb, Ph.D. and Joe F. Pedelty, began with a sample 100,000 sq.ft. building in New York City that spent $239,904 on ventilation each year. There were 667 employees in the building, each earning an average salary of $30,000, for a total of $20.01 million per year. The usual absentee rate was 2.6 percent, at an annual cost of $520,260. Half of that was attributed to upper respiratory illness, with 27 percent of the respiratory illness cases attributed to IAQ-related problems.

Holcomb and Pedelty increased the ventilation rate from 5 cfm per person to 20 cfm per person. They estimated an initial cost of $28,500 and additional operating costs of $5,575.70. Costs to increase ventilation, of course, will vary by building. At the conclusion of the study, the anticipated savings due to decreased absenteeism was $70,235—a payback of about six months and a net savings of nearly $65,000 per year. Calculations related to absenteeism and its causes were obtained from The National Center for Health Statistics.

Holcomb and Pedelty presented these findings:

> **Studies attempting to correlate increased ventilation rates with improved IAQ or a reduction in [sick building] symptoms have obtained mixed results ... [however,] most data generated in IAQ studies suggest an association between ventilation rates, IAQ, [sick building syndrome] symptoms and employee productivity and absenteeism ... [and] increasing ventilation rates may potentially provide substantial savings to an employer.**

On a larger scale, a study by the National Energy Management Institute (NEMI) concluded that productivity could rise by as much as $54.5 billion a year nationwide if all commercial building owners adhered to existing voluntary indoor ventilation and temperature guidelines (ASHRAE Standard 62-1989 and ASHRAE Standard 55-1981, respectively) (see Table 1-1). The study claims that productivity gains alone within two years would recoup all necessary capital improvement costs for implementing ASHRAE IAQ standards. In fact, according to NEMI, net productivity gains would outstrip costs by 700 percent over 20 years

NEMI estimated the total value of productivity could be improved by 3.5 percent, mainly through reduced employer-paid medical costs, increased profits from more productive employees, reduced absenteeism and increased product quality. The study projected annual medical costs would fall by $435 million. According to NEMI, the improvements to many buildings would have a payback of less than seven months, while the average payback from productivity benefits is an estimated 1.6 years. In addition,

Table 1-1. NEMI projections of productivity gains due to adhering to existing voluntary ventilation and temperature guidelines.

Inventory	
Number of commercial buildings in U.S.	4.5 million
Total space (excluding warehouses/parking)	47.9 billion sq.ft.
Number of workers	68.3 million
Productivity Benefits	
Annual total	$54.5 billion
Health benefits	$435 million
TOTAL from above	$54.9 billion
General productivity benefits	$805/worker/yr
General productivity benefits	$1.15/sq.ft.
Cost To Implement	
All IAQ improvements	$88.6 billion
Average cost/sq.ft.	$1.85
Average cost/worker	$1,297
Initial average payback	1.6 yrs
Annual cost to sustain	$4.8 billion
Net 20-yr value benefits minus cost	
Per sq.ft. for all improvements	$10.02/sq.ft.
Per worker for improvements	$7,025/worker

the study stated: "Advances in heat recovery technology and new HVAC systems will permit many buildings to actually realize a decrease in energy costs."

Marketing

Avoiding lawsuits and other IAQ problems is instrumental in maintaining a company's positive public image. Similarly, there may be marketing advantages for facility managers who maintain a rental property with acceptable IAQ. Tenants are more likely to look more closely at a "green" building—where wastes are kept to a minimum or recycled, where the HVAC system is efficient, where the indoor air is clean and free of contaminants.

A Word of Caution - These benefits can be promoted, but use caution. Some filter manufacturers and other clean air products have come under fire from consumer advertising councils for exaggerating their claims. "Environmentally safe" or "environmentally friendly" may be deceptive if the product, or in this case the building, does not really offer any environmental benefit. Would not the environment be better off if there were no building there at all? "Non-toxic" and "biodegradable" are words that have come under scrutiny—most products are neither. The U.S. EPA defines some of these terms in its report, *The Evaluation of Environmental Marketing Terms in The United States.*

Rather than getting into cases of exact wording, simply use restraint. It is proper to say that the building uses chlorofluorocarbon (CFC)-free air conditioning, if it does, or that the air conditioning system will not harm the ozone layer. Or that steps have been taken to clean the air with better filtration or with a smoking ban. But do not try to ensure a "healthy, safe environment." Lawyers will jump all over broad, vague terms such as that. Is it a healthy building if someone slips and falls on a freshly washed floor? Maybe not, no matter how much money has been put into the building's mechanical system.

GOVERNMENT ACTIVITY ON IAQ

Many of the indoor air programs at this point are voluntary. As of this writing, several national IAQ laws have been proposed, but none has passed. Speculation is meaningless, given the constantly shifting political landscape. National IAQ bills have come and gone; regulations can be proposed one day and permanently shelved the next. Many local laws take precedence, as do building codes or other restrictions.

National laws on IAQ in some form are no doubt coming. As soon as the regulatory powers that be can agree on definitions, content and purpose, then fight to gain the necessary approval of Congress and the Executive Branch.

OSHA

On the aftermath of the Republican takeover of the Congress in 1994, Joseph A. Dear, head of the Occupational Safety and

Health Administration (OSHA), was quoted as saying, "On election night, I slept like a baby. I woke up every two hours and cried." OSHA had developed its own set of IAQ guidelines for the workplace, and as of this writing was conducting lengthy hearings on the matter before implementation could begin. The process may take years, and whether there will actually be new laws is debatable, given the current anti-regulatory mood of the Republican majority. Nevertheless, facility managers will be called on to "do the right thing," whether for legislative, regulatory or other reasons. The trend is definitely toward stricter compliance.

As a minimum, if a law is passed, expect OSHA to require:

- A brief description of the building and its mechanical system(s).

- Schematics and/or as-built construction documents.

- Operating instructions so the equipment will perform within design criteria.

- Performance criteria, such as ventilation rates and the acceptable range for temperature and humidity.

- Any plans for modifications if conditions or occupancy change.

- A description of the equipment, type and capacity, with maintenance practices and intervals on separate checklists.

- Maintenance manuals and commissioning documents.

    OSHA's proposed rules also require:

- A ban on environmental tobacco smoke (ETS) indoors, or requirement of a dedicated special smoking area.

- A designated IAQ coordinator at the facility.

- Training for staff on IAQ and IAQ procedures, in addition to personal protective equipment.

- The development of a written IAQ plan, including an inspection checklist. (Refer to Appendix III.)

- Record-keeping for IAQ complaints, training materials, operating & maintenance manuals and other items described above.

- Longer hours for HVAC operators.

- Provide maximum ventilation rate required by code or occupancy.

- Distribute information to building occupants regarding IAQ.

- Control trigger points for levels of carbon dioxide and relative humidity.

OSHA estimates the cost of compliance at $8.8 billion nationally, most of it attributable to documentation, inspection and system maintenance. Assuming the maintenance is already being done or should be, the major hurdle will be in record-keeping. Some of this task can be automated, using direct digital controls or a building automation system. Do not look at this as one more expensive layer of government interference. Instead, look at it as an opportunity to save energy, maximize assets and create a paper trail to gain protection from legal threats. For detailed information on dealing with OSHA and an OSHA inspection, see Joseph Gustin's book, *Safety Management: A Guide For Facility Managers*, a companion volume in the Facilities Management Library.

Codes
National and local building codes are beginning to address the IAQ issue as well, both for commercial as well as residential buildings, sometimes implementing ASHRAE Standard 62-1989 for ventilation. As we will discuss in Chapter 4, as of the time of writing Standard 62-1989 was in the process of revision with a targeted release date in a 1996-1997 time frame, and the results from Standard 62-2001.

U.S. EPA's Building Air Quality Alliance
The Building Air Quality Alliance was developed as part of the U.S. EPA's "carrot and stick approach," a way of getting the

private sector proactive about IAQ without adding to current regulations and restrictions. According to U.S. EPA, the Alliance was developed collaboratively by a number of individuals representing organizations who are committed to improving IAQ practices. It was based on the concept that a voluntary partnership of governmental, non-governmental organizations and the private sector can more readily achieve substantial results in improving IAQ in buildings than any one group can alone. The U.S. EPA hoped so many would voluntarily comply with such a program that it could eventually become standard operating procedure in most office buildings.

The guiding principles of the Building Air Quality Alliance call for building owners and facility managers to:

- Prioritize good IAQ.

- Understand the sources of indoor pollution and how to solve the problems.

- Establish a program and practices that maximize good IAQ.

- Quickly resolve discovered problems.

- Communicate effectively, openly and honestly on IAQ issues with building occupants.

Although it ran into a bureaucratic and legislative buzzsaw before being at least temporarily sidelined in 1995, the Building Air Quality Alliance was established to work like this:

Owners of lease properties who certify that a series of steps has been followed in their buildings to ensure healthy indoor air, and can document these steps, will win a sort of national seal of approval.

They may then use this in future marketing programs or in contract language to attract or retain tenants, and possibly even as a legal defense should a problem (or lawsuit) ever arise. School boards would be eligible to apply.

The concept is still a useful one. Two key groups were to comprise the Alliance: Alliance members and building partners.

Table 1-2. Existing criteria for lAQ. Source: Healthy Buildings and the U.S. Department of Commerce.

Parameter	Criteria	Reference
Outdoor Airflow Rate	10 L/s (20 cfm) per person	ASHRAE 62-1989, Table 2
	2.5 L/s (5 cfm) per person	BOCA National Mechanical Code 1990, Table M- 1603.2-1
Pressure Relationships between Zones	Restrooms mechanically exhausted with no recirculation.	ASHRAE 62-1989, Table 2 BOCA National Mechanical Code 1990, Section M-1603.4
	Restroom design exhaust capacity shall be greater than or equal to the design supply capacity	BOCA National Mechanical Code 1990, Section M-1602.9.3
Carbon Dioxide	1000ppm	ASHRAE 62-1989 [6], Table 3
Carbon Monoxide	9 ppm, 8-hour average 35 ppm, 1-hour average	EPA - National Ambient Air Quality Standards [12] (ASHRAE 62-1989, Table C- 1)
	Less than 9.6 ppm, limited or no concern Greater than 26 ppm, concentration of concern	WHO [13] - Continuous exposure (ASHRAE 62-1989, Table C-4)
	17 ppm, annual average 26 ppm, 24-hour average	BOCA National Mechanical Code 1990 [5], Table M-1603.2.1
Formaldehyde	0.40 ppm. target level for homes	HUD - standard for manufactured homes [14] (ASHRAE 62-1989 Table C-1)
	0-05 ppm, limited or no concern 0.10 ppm, concentration of concern	WHO - Long and short-term exposure (ASHRAE 62-1989 Table C4)
Particulates	50 µg/m^3, annual average (PM 10) 150 µg/m^3, 24-hour average (PM 10)	EPA - National Ambient Air Quality Standards (ASHRAE 62-1989 Table 1)
	60 µg/m^3, annual average 150 µg/m^3, 24-hour average	BOCA National Mechanical Code 1990 [5], Table M- 1603.2.1
Radon	Action level for homes: 4 pCi/L (picocuries per liter)	EPA 1988 - Radon Reduction Techniques for Detached Houses, Technical Guidance. (ASHRAE 62-1989 Table 3)
Total Volatile Organic Compounds (TVOC)	Comfort range; less than 200 µg/m^3 Multifactorial exposure range: 200 - 3000 µg/m^3 Discomfort range: 3000 - 25000 µg/m^3 Toxic exposure range: greater than or equal to 25000 µg/m^3	Mølhave, Indoor Air '91 [15] This is not a standard, but is based on a combination of field and controlled climate chamber tests.
Thermal Comfort	Predicted Percent Dissatisfied (PPD) less than or equal to 10%	ASHRAE Standard 55-1992 [16] ISO Standard 7730 [17] Recommendation given in Annex A which is not part of the standard

Members would include national non-profit organizations, associations or government entities. Building partners would include building owners or their designees. Their goal: By drawing diverse interests together to work collaboratively, the Alliance can make a difference and improve indoor air quality.

To learn more about this program, contact the U.S. EPA via its IAQ hotline for a status report on this program: (800) 438-4318.

THE ROLE OF THE FACILITY MANAGER

A sign of the times: IAQ complaints have skyrocketed in recent years. The National Institute for Occupational Safety and Health (NIOSH) reported that it received more than 7,000 phone requests and 760 written requests for help from employers during the '93 fiscal year versus about 150 on average for previous years. That is a fivefold increase.

The responsibility for identifying and solving IAQ problems often falls to the facility manager, as he is the professional responsible for successfully integrating the occupant with his physical environment. Figures 1-8 through 1-11, which show the results of a recent IFMA survey on IAQ, illustrate the growing awareness of this responsibility, and a growing commitment to good IAQ.

We must also note that the role of facility manager has changed. It has gone from caretaker to more broad-based, sophisticated management practices with the task of maintaining and improving the value of the building as an asset for its owner.

Building owners are looking for facilities managers and man-

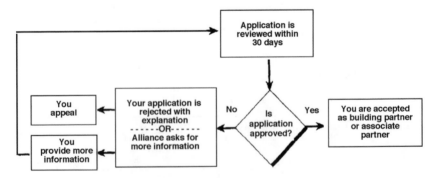

Figure 1-6. The U.S. EPA's Building Air Quality Alliance, although temporarily sidelined by budget wrangling in 1995, is a useful concept that encourages and rewards voluntary programs to achieve good IAQ. Shown is the process for becoming a participant in the program. Courtesy: U.S. EPA.

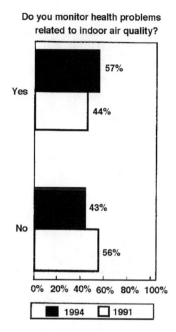

Do you monitor health problems related to indoor air quality?

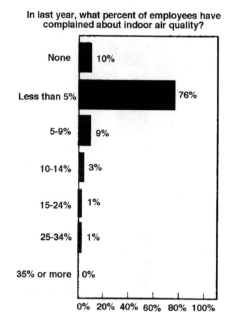

In last year, what percent of employees have complained about indoor air quality?

Figure 1-7. 1994 IFMA IAQ survey. Courtesy: IFMA.

Figure 1-8. 1994 IFMA IAQ survey. Courtesy: IFMA.

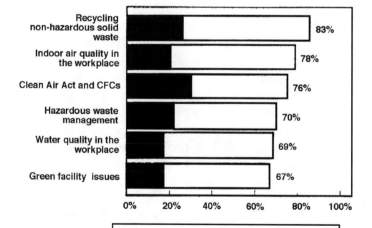

How would you rate your knowledge of:

Figure 1-9. 1994 IFMA IAQ survey. Courtesy: IFMA.

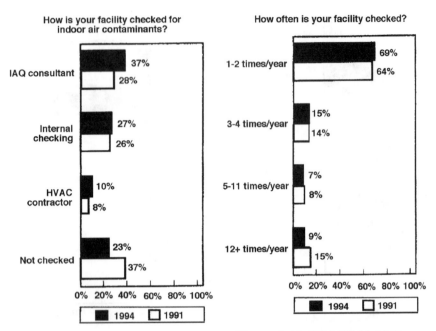

Figure 1-10. 1994 IFMA IAQ survey. Courtesy: IFMA.

Figure 1-11. 1994 IFMA IAQ survey. Courtesy: IFMA.

agement companies that understand the issues, which includes the investment, physical and operating needs of the facilities they are serving. Only in that way can one maximize the value of an asset as fixed—and enormous—as, for example, an office building. "You can't just walk in to a client and say, 'We now offer facilities management,'" is how one general manager put it.

Corporate facilities manager Charles Rickett makes an observation that no doubt many in this profession can second: "My personal gut reaction to the IAQ phenomenon is that it is real; like everything else in facilities management, perception is reality."

Remember the haunted house that nobody was willing to walk by when we were kids? Was it really haunted? Did it matter?

Consider one early case of sick building syndrome involving a school in Hagerstown, MD, Northern Middle School. Almost immediately after the school was occupied in 1980, complaints arose: poor temperature control, inadequate ventilation, odors, headaches, etc. Early remediative efforts failed, and a crisis situation arose. A number of changes to the building were made, involving the mechanical system controls, air ducts, air intakes and other HVAC components. The problems were solved. Looking back, researchers concluded: "If complaints go unheeded, or if occupants of a building are not convinced that corrective action is being taken, the situation is bound to become worse."

Agreed: doing something is better than doing nothing. All right then, where do we begin?

Solving an IAQ problem can be as simple as closing windows that overlook a freeway to shut out tailpipe pollution and noise, to upgrading the HVAC system. The solutions are not always complex, but there are many possible solutions because there are many conditions that can contribute to an IAQ problem. In this book, we will take a broad-based approach that covers modifying the HVAC system and other measures. To tackle the full range of IAQ problems, we must ensure:

- Recognition and knowledge of IAQ hazards (Chapter 2).

- Adequate ventilation (Chapters 3 and 4).

- Adequate temperature (Chapters 3 and 6).

- Adequate levels of humidity, as in proper humidification and dehumidification (Chapter 6).

- Adequate air balance and pressurization (Chapters 3 and 4).

- Adequate filtration (Chapter 5).

- Scheduled maintenance that can solve current problems and prevent future problems (Chapter 7).

- A comprehensive on mold (Chapter 9).

- That the HVAC system operates as designed via commissioning (Chapter 3).

- A comprehensive and effective IAQ program (Chapter 8) that includes, among other things:

    1. Taking IAQ-related health complaints seriously with documentation and response.

2. Building an IAQ team, establishing roles and determining an IAQ policy.

3. An investigative process of surveying (including on-site inspection, reviewing complaint records and possibly conducting interviews) a building to identify IAQ problems. The process should include identifying 1) the contaminant, 2) the source of the contaminant and 3) its pathway to people. Example: People on the east side of the building complain of headaches early in the morning and late in the afternoon (problem). After an investigation, it is discovered that tailpipe pollution (contaminant) is coming from a nearby freeway during rush hour (source) and entering through open windows (pathway).

4. Mitigation of the problem now and in the future.

Taking the above steps should identify, solve and prevent basic IAQ problems, thereby reaping the benefits of good IAQ and avoiding penalties such as lawsuits. Finding and solving IAQ problems requires a bit of detective work and other expertise, so the facility manager may find it desirable to hire a qualified consultant to help.

Sources
Andersson, K., Norlen, U., Ph.D., Fagerlund, I., Hogberg, H., and Larsson, B., "Domestic Indoor Climate in Sweden: Results of a Postal Questionnaire Survey" ASHRAE paper, 1992.

Bernheim, Anthony; and Black, Dr. Marilyn; "Green Architecture: Indoor Air Quality Design is Cost Effective," paper, American Institute of Architects National Convention, May 1995.

Building Air Quality: A Guide for Building Owners and Facility Managers Handbook, Building Owners and Managers Association (BOMA), p. 19.

Environmental Issues in The Workplace II, International Facility Management Association (IFMA), 1994.

Fedrizzi, S. Richard, Director of Environmental Marketing for Carrier Corp., and chairman of the U.S. Green Building Council, *Skylines*, January 1995.

Fedrizzi, S. Richard, "Going Green: The Advent of Better Buildings," *ASHRAE Journal*, December 1995.

Gallo, Francis M., "Designing a Proactive Indoor Air Quality Program," *Skylines*, January 1995, p. 18.

Ganick, Nicholas, PE, Ronald V. Gobbell, AIA and Hays, Steve M., PE, CIH, *Indoor Air Quality: Solutions And Strategies*, New York, McGraw-Hill, 1995.

Goldman, Ralph F., Ph.D., "Productivity in the United States: A Question of Capacity or Motivation?" ASHRAE paper, 1994.

Halliwell, Jack L., PE; "IAQ Diagnostics for Building Owners, Managers And Their Consultants," paper at Healthy Buildings '95 Conference, Chicago, sponsored by the National Coalition on Indoor Air Quality (NCIAQ).

Hansen, Shirley, Ph.D., *Managing Indoor Air Quality*, Atlanta: The Fairmont Press, 1991.

Hartkopf, Volker and Loftness, Vivian, Carnegie Mellon University, Pittsburgh, "Innovative Workplaces: Current Trends and Future Prospects" paper presented at the 1992 IFMA Conference.

Hedge, A., Burge, P.S., Robertson, A.S., Wilson, S., Wilson and Harris-Bass, J., "Work-Related Illness in Offices: A Proposed Model of The 'Sick Building Syndrome,'" Pergamon Press, Environment International, Vol. 15, No. 1-6, 1989.

Helsing, Knud J., Billings, Charles E., Conde, Jose and Giffin, Ralph, "Cure of a Sick Building: A Case Study" Environment International, Pergamon Press, Vol. 15, No. 1-6, 1989.

Hennessey, John F. III, "How to Solve Indoor Air Quality Problems," *Building Operating Management*, July 1992.

Holcomb, Larry C., Ph.D. and Pedelty, Joe F., "Comparison of Employee Upper Respiratory Absenteeism Costs with Costs Associated with Improved Ventilation," ASHRAE paper, 1994.

Holohan, Dan, "Read Any Old Books Lately?" *PM Engineer*, Business News Publishing Co., February/March 1995.

Indoor Air Quality Update, various issues, Cutter Information Corp., Arlingon, MA.

Int-Hout, Dan, "Total Environmental Quality," ASHRAE paper, 1994.

Indoor Air Quality Manual, Sheet Metal and Air Conditioning Contractors National Association Inc., 1988, third printing 1990.

Introduction to Indoor Air Quality: A Reference Manual. Washington, DC: U.S. Environmental Protection Agency, July 1991.

"Keeping HVAC Systems Safe," *Maintenance Technology*, September 1992.

Kirsch, Lawrence, "Liability for Indoor Air Pollution," paper from a symposium at the 198th national meeting of the American Chemical Society, 1989.

Lord, D., Ph.D., "Air Quality in Western Culture: A Short History," ASHRAE paper, 1995.

Lorsch, Harold, Ph.D., PE and Abdou, Ossama, Arch.D., "The Impact of the Building Indoor Environment on Occupant Productivity," ASHRAE paper, 1994.

Manko, Joseph, "Investing a Few $$ Can Avert IAQ Legislation," *Econ*, January 1993.

Milam, Joseph A., PE, "A Holistic Approach to Improving Indoor Environmental Quality," Designing Healthy Buildings Conference, The American Institute of Architects.

Mintz, Alan; "IAQ: A Business Opportunity for Contractors," from a talk given at Air Conditioning Contractors of America 26th annual meeting, New Orleans, 1994.

Pomeroy, Christopher D., *Green Building Rating Systems: Recommendations for a United States Rating System*, revised February 17, 1995.

"Productivity And Indoor Environmental Quality Study," National Energy Management Institute, prepared by Dorgan Associates, WI, 1993.

"Utility Proves Old Buildings Need Never Die," *Green Building Report*, quarterly publication of the U.S. Green Building Council, January 1995, p. 4.

Wheeler, Arthur E., PE and Olcert, Robert B., Ph.D., CIH, OSD, Indoor Air Quality course, ASHRAE 1991, printed course material.

Chapter 2

IAQ Hazards to Human Health and Productivity

Poor indoor air quality can make us seriously ill, even kill us. But more often, it is a nuisance factor that masks an assortment of various less-threatening but still troublesome maladies leading to loss of work days, reduced productivity, unhappy tenants, broken leases, vacant properties, strained management-owner relations, etc. Common materials can affect us in bothersome ways. All of this is a part of building IAQ.

ASTHMA AND ALLERGIES

If you are one of those who suffer from asthma or allergies, you are not alone. Between 1982 and 1991 the number of asthma cases reported in the United States increased by nearly 50 percent, according to the American Lung Association. In 1991, more than 5,000 Americans died from asthma, with some 200,000 emergency room treatments administered and an overall estimated $6.2 billion spent for various remedies and temporary relief. Allergies, which often lead to asthma, are responsible for another 2,000 deaths according to the American College of Allergy and Immunology. According to the U.S. Centers for Disease Control and Prevention, almost one in 20 Americans now suffers from this respiratory condition. Allergies and asthma cause more than 130 million lost school days and 13.5 million lost work days each year. Children are particularly at risk from asthma, because their defense mechanisms are not as established.

Eighteen million Americans are allergic to the fecal matter and dead body parts of the common dust mite (10 of which can fit in the period of this sentence). Dust mites are the number one allergen found indoors, followed by cats and dogs, then cockroaches and mold. Unfortunately, there is no way to avoid dust mites, as they are part of our environment. They are not always a problem, however. For one thing, they have to become airborne. Simple cleaning, such as vacuuming the carpeting, is often enough to remove them—although this also helps to make them airborne, ready to spring safely into new habitats.

"Hay fever" is not caused by hay and it is not a fever. Its technical term is allergic rhinitis, which is referred to in many reports of sick building syndrome or problem buildings. It is caused by pollen from trees, grasses, weeds or mold, and affects one out of five people. Allergies tend to run in families, and "hay fever season" can actually include spring, summer or fall depending on geographic location. They cannot be avoided by going to Arizona. Not since so many emigrants took their beloved transplants, flowers and grass there with them.

A study in Sweden reported that one-third of the population there has or has had some form of allergy or hypersensitivity, and the numbers are on the rise. It blamed such pollutants as tobacco smoke and mold, as well as changes in households including more use of carpeting, gaseous agents such as formaldehyde, NO^2 and ozone, and less air circulation. Said the study: "Allergen levels in public premises such as schools and day nurseries are often sufficient for sensitization and for acute symptoms in those already sensitized." Even this change in housekeeping was not overlooked: "A change in housing hygiene which has taken place over the past few decades and which is presumably an essential change, is that bedclothes are less often aired in windows." Giving rise to increased populations of dust mites.

Linda Bean, a mother in Maine, was puzzled when her young son Troy was hospitalized seven times in one year, missing 47 days of school (according to a Honeywell Inc. case study). Mysteriously, the symptoms cleared up when he was at home—sick of school? Perhaps. It turns out the fourth grader was discovered to be especially sensitive to mold, mildew and dust, which happened to be prevalent in the portable classroom building he attended. His illness was attributed to asthma.

Particulate matter is a contributor to asthma. Ozone, too, and certain biological agents such as bacteria, fungi and animal dander. Asthma used to be associated only with pollen, spores and mold, but now is thought to be part of an overall decline in urban air quality. We often cannot get away from elevated levels of these things outdoors, but discomfort and illness indoors is almost always influenced by building factors. Filtration can eliminate some of these factors, but can also contribute by themselves if they are not changed on a regular basis. A build-up of dirt and captured particles can serve as a medium or magnet to attract even more of this material, allowing it to build up and then disseminating it into the air stream by recirculating it. Conditions worsen if this material is allowed to become damp or wet, becoming a suitable breeding ground for microbes and bacteriologicals.

> **Allergies and asthma cause more than 130 million lost school days and 13.5 million lost work days each year.**

Exposure Can Induce Symptoms

Several different types or symptoms of illness can come from being exposed to allergens, including skin reactions, sneezing, labored breathing and other problems. These vary according to person, probably because of differences in immune systems, and are thought to be at least to some extent genetic. An estimated 20 percent of the population has some predisposition to allergies. Allergic asthma affects some 3-5 percent of the population, and is accompanied by narrowed airways and production of mucus, which also blocks airways. This differs from allergic rhinitis (hay fever) and is usually alleviated as soon as the source, usually pollen, is removed. Conditions can worsen if bacterial infections follow. A more serious condition is hypersensitivity pneumonitis, once thought to be limited to agricultural and certain industrial workers. It is now known that some people can become hypersensitized while working in offices, just by exposure to certain fungi. Estimates of pneumonitis occurring in offices ranges from 1-4 percent.

"In the last five years, fungi have come to be seen as quantitatively the most important bioaerosols with respect to health in indoor air," stated a report from the United States National Academy of Sciences. The report devoted perhaps 15 percent of the text to information about eliminating fungal growth in buildings. "There is consensus that it is harmful to health if fungi grow to such an extent that their spores and volatiles become appreciable contaminants of indoor air," concluded the report.

The Fort Worth, TX-based duct-cleaning company Clean-Aire Technologies commissioned an independent study in 1989 by Mycotech Biological Inc., to sample eight test houses before and after air duct cleaning. "We knew we were making people feel better, but we wanted to know that our service was valid," said Bob Allen. Studies of any sort in this area were lacking. This particular study revealed a reduction of more than 90 percent in fungus spores and airborne mold. The study, according to Allen, validated anecdotal evidence provided by customers. It was listed as recommended reading by the American College of Allergy, Asthma and Immunology, and included in the *Yearbook of Allergy*.

ASBESTOS

"The asbestos of the '90s ..."

"Let's hope this isn't another asbestos ..."

Public reaction to environmental threats has gone full circle. At first, we rightfully were outraged at heretofore unseen threats such as DDT. Today, consumers are deluged with many real and perceived threats, from fat and cholesterol in their diets to drunk drivers, unsafe products, stress, air pollution, lead, food additives, etc.

Were we wrong on asbestos? Asbestos is a known carcinogen that was used generously in many common everyday products, including some 3,000 that are found in most homes. The U.S. EPA estimates that 3,000-12,000 cancer cases a year can be traced to asbestos. It is still found in many older buildings in the form of ceiling tiles, floor tiles, pipe wraps, etc. No, we were not wrong to curb the use of this substance, but it is generally agreed that our actions on asbestos removal were perhaps an over-reaction. Millions of dollars were spent in emergency funding to remove asbestos from buildings where it would probably would have done little or no harm.

Lawsuits continue to be filed to this day by people who once worked with asbestos. An Asbestos Victims Special Fund Trust in Philadelphia continues to pay out damages. Tens of millions of dollars have been paid so far in settlements. The statute of limitations on filing a lawsuit does not begin to run until a person is diagnosed with an asbestos-related disease or until they should have known they have an asbestos-related illness. It can take 20 years or more from the time one is exposed to the dust before the onset of symptoms of the disease.

OSHA Rules on Asbestos

Rules on asbestos are constantly changing; one set of revised standards was issued by OSHA on October 11, 1994. It is more rigorous than the standard it replaced. A time-weighted average for exposure limits to fibers has been changed from 0.2 to 0.1 fibers greater than 5 micrometers in length per cubic centimeter of contaminated air. More stringent respirator standards were included in this new standard along with other revisions. For one thing, the revisions *assume* asbestos to be present in buildings constructed before 1981, unless it can be proven otherwise, calling for various actions to be taken in response. Certain courses of action concerning worker protection *must* be undertaken for activities where the material could be disturbed, such as during remodeling and maintenance activities—unless a building owner can prove that asbestos is *not* present, which is sometimes a difficult task. This proof must come from testing or live inspection, not just based on building records.

> The U.S. EPA estimates that 3,000-12,000 cancer cases a year can be traced to asbestos.

In addition, anyone who comes into contact with asbestos, even a building maintenance worker, must have some form of special training, according to OSHA. Two hours of "awareness training" for custodians and up to 16 hours of training for those involved in repair work is required. Note: a pre-1981 building is *assumed* to contain asbestos unless it can be proven otherwise.

Encapsulation

Encapsulation of asbestos is an option in some cases. This means a coating or shielding material is placed over it to catch any stray fibers that might become airborne. In other cases, it is best to leave the material alone, and in place. The worst thing is to start in on a do-it-yourself asbestos-removal job. Vacuum cleaners only serve to push these fine particles into the air, where they can remain airborne for long periods of time. If asbestos must be removed, hire licensed, insured professionals who will use special tools and equipment to safely do the job.

For More Information ...

The federal budget for asbestos-funding within the EPA was slashed for 1996, with efforts being transferred to the states. For more information, the U.S. EPA has a publication, _Managing Asbestos in Place: A Building Owner's Guide_, which can be ordered by calling (202) 554-1404.

FORMALDEHYDE AND VOCS

According to a survey of environmental factors in the workplace conducted by IFMA, fully half of those surveyed are now monitoring for formaldehyde and other volatile organic compounds (VOCs) in the workplace. This is up from only 11-12 percent reported in a similar survey taken three years earlier.

Formaldehyde (HCHO) is a flammable, colorless gas with a characteristic odor found in some form in eight percent of all of the products we use. In 1895, it was being widely used as an inexpensive fumigant for disinfection. Homes harboring an infectious disease were routinely fumigated with formaldehyde gas, a practice which lasted for decades. Its odor threshold is about 1 ppm, but some sensitive individuals can detect it to as low as 0.05 ppm. In sufficient quantities it can cause skin, respiratory, throat and eye irritation; headaches, fatigue, nausea, dizziness, breathlessness and problems with concentration and memory. It has been reported, but not proven, to be associated with altered reproductive function in women. Repeated contact with low concentrations can cause sensitization to future exposures, according to the EPA, which has also labeled it a carcinogenic risk to humans.

Formaldehyde can be coated with paint that contains a vapor retardant. Some people, however, are especially sensitive to formaldehyde.

Formaldehyde material is most prevalent in insulation, hardwood plywood, particleboard and the fiberboard found in countertops, furniture and cabinets (see Table 2-1). Manufacturers are using more of those materials while trying to use less formaldehyde, and in older buildings much of the formaldehyde has already off-gassed so that it should no longer present a problem.

Table 2-1. Potential indoor sources of formaldehyde.

Pressed Wood Products	Hardwood plywood, particle board, medium density fiberboard, decorative paneling
Insulation	Urea-formaldehyde foam insulation, fiberglass made with HCHO binders
Combustion Sources	Natural gas, kerosene, tobacco, auto exhaust
Paper Products	Grocery bags, waxed paper, facial tissues, paper towels, disposable sanitary products
Stiffeners, Fabric Treatments	Floor coverings, carpet backings, adhesive binders, fire retardants, permanent press fabrics
Others	Plastics, cosmetics, deodorants, shampoos, disinfectants, laminates, paints, ink, fertilizer, fungicides

Presence of VOCs is usually stated as TVOC, for total volatile organic compounds. Ten of the most common are toluene; benzene; ethylbenzene; tetrachloroethylene; 1,1,1 trichoroethane; styrene; limonene; isopropanol; ethanol; and xylenes. VOC sensors are available, but they have limitations, including sensitivity and accuracy—nor are there presently guidelines in place to describe what levels of VOCs are safe or unsafe. One Johnson Controls study that monitored 240 spaces in 15 buildings, summer and winter, determined no correlation between a VOC sensor's readings and that of a trained human sensory panel.

Labeling

Manufacturers of office furniture must by law attach labels to their products if they off-gas more than a certain amount of formaldehyde. Products that release this substance at less than 0.1 ppm need not have a label, which is exactly what Grand Rapid, MI-based office product manufacturer Steelcase sought: "label-free" status for its products. Any products which off-gas more than 0.5 ppm of formaldehyde must carry another label: *Potential Cancer Hazard*. (It is hard to imagine a bigger turn-off when buying a product to see that label.)

Steelcase has done a study, Formaldehyde Emission Characteristics of Finished Wood Furnishings and Individual Construction Materials. As a result, it claims it is better able to manufacture, design and market wood furnishings that have low formaldehyde emissions and a low impact on the indoor environment. It noted wide variations in the amounts emitted. Use of domestic versus imported particle board and use of a sealer were factors (imports do not have to meet HUD and ASTM test standards). More often, type of topcoat used varied the amount of formaldehyde emitted. Among the conclusions: while particle board was identified as a contributor of formaldehyde, it was also identified as an inhibitor—it can act to dampen the impact of the topcoats used. In addition, it was able to work with suppliers to modify and substitute raw materials to produce a low formaldehyde-emitting wood product. Steelcase and Air Quality Sciences Inc. together developed a system for ranking office furniture according to VOC and formaldehyde emissions as high, moderate or low based on results from an environmental test chamber under constant conditions of 23°C, 50 percent relative humidity and one air change per hour. AQS compiled data from more than 500 indoor air surveys of air in homes, schools and commercial buildings. The data showed that cleaning compounds account for numerous airborne VOCs and that many of these VOCs may be irritants and odorants.

The Medite Corporation of Medford, OR, markets its "Medex" and "Medite II" formaldehyde-free plywood and wood building products—wood naturally contains trace amounts of formaldehyde, but no additional amounts are added to these products. The availability of products ranked according to a sort of IAQ index is increasing: the Carpet and Rug Institute has its own emissions criteria program.

In Germany, there is a "Blue Angel" program which focuses on the environmental acceptability of products based on several factors, including chemical emissions, recyclability, waste generation, noise and hazardous substances.

> In sufficient quantities, formaldehyde can cause skin, respiratory, throat and eye irritation; headaches; fatigue; nausea; dizziness; breathlessness; and problems with concentration and memory. Repeated contact with low concentrations can cause sensitization to future exposures, according to the EPA, which has also labeled it a possible carcinogen.

RADON

Remember radon? Many people do not. Radon is a cancer-causing radioactive gas, unseen and unsmelled. Years ago, there were public service announcements warning of the possible dangers of radon. Many people dismissed them: Certainly, it seemed as if there are enough environmental problems and cautions out there, and the outcry on radon often takes a back seat to others that are more visible, more attention-getting in the public eye.

True, there have been many environmental scares that have not panned out. This is where some of the reluctance comes in accepting indoor air quality, for one, as a full-blown environmental priority. Apples and alar come to mind. Caffeine in our coffee, alcohol, partially hydrogenated vegetable oil, sulfites and meat that is too rare or too well done over a charcoal fire. Asbestos, too. And radon. People are tired of hearing "Wolf!," and largely want to be informed about one more daily threat only when it is knocking on their door. In other words, when it is an acknowledged, serious health hazard: capable of killing Uncle Henry or their next door neighbor, and not one that merely provides more grist for the television news programs or reports.

This is understandable. Many large-scale changes, even the movement away from CFC-type refrigerants, are fraught with hazard. Alternative refrigerants can be expensive—some equipment must be overhauled or replaced completely. Worse, there are always lurking hazards that we may not even be aware of—despite rigorous testing for flammability and exposure to humans, suppose one of the alternative refrigerants turns out, somewhere down the road, to be as dangerous or at least as environmentally unacceptable, as CFCs? It is a risk we have to live with.

Our Greatest Radiation Risk

Yet few would disagree that radon is an environmental risk. There is a large body of evidence that says radon gas—or, more accurately, radioactive "daughters" that are its by-products—in large portions can be deadly, certainly a contributor to lung cancer. Radon is by far the highest radiation danger that the American public faces, according to the U.S. EPA. While radon has a half-life of only 3.8 days and would be exhaled before doing any harm, radon decay products such as bismuth, polonium and lead can emit alpha particle radiation quickly after being inhaled.

The Watras family of Boyertown, PA, were the recipients of some bad news in 1984. What happened to them focused attention on the radon problem and brought it to the attention of many. Stanley Watras worked as a construction engineer at a nuclear powerplant. He began to set off the radiation detectors there, even though his job did not place him in the vicinity of any radioactive materials. Perplexed investigators finally tested his house when they could not pinpoint the cause of the radioactivity, and found radon levels a thousand times higher than the U.S. EPA's 4 picocuries per liter action level. The family quickly moved out of the house and remediation was begun, fortunately paid for by the local utility company as part of a research project. The radon was traced to a uranium-containing block of granite underneath the house. Some neighboring homes tested got a clean bill of health. The final tab on the Watras home was $32,000, after which they moved back in. They were safe once more, but no one can guess the effects of the radon gas they inhaled before the problem was discovered.

How did radon enter the home? There are three main ways for radon to enter buildings: transport through soil via cracks or openings around the foundation; emanation from the very building materials we use such as concrete or rock; and transport via water and natural gas. Soil is the most common source—whereas radon from building materials is generally isolated to particular locations, even particular buildings.

Radon is measured in picocuries: a picocurie is 1/trillionth of a curie, a unit of measure for radioactivity, named of course for Marie and Pierre Curie, early pioneers in the study of radioactivity. The present permissible limit is 4 pCi/l—some say that the risk of developing lung cancer from this amount of radioactivity is the same as the risk of dying in an auto accident. The jury is still out on exposure levels: The State of Iowa, for example, is said to have some of the highest levels of radon, yet has a low per capita rate of lung cancer. Just as confusing: A high reading in a neighbor's home does not mean your home has one also; nor does the opposite hold true.

Thousands of Preventable Deaths

The U.S. EPA asserts that radon is the cause of 20,000 cases of preventable lung cancer deaths every year; that radon testing is both accurate and reliable; and that radon problems can be fixed. More deaths are caused annually by exposure to radon gas, according to the U.S. EPA, than by drowning, fires or airline crashes—7,000-30,000 deaths per year. There is no way to avoid this radiation completely. Fifty-five percent of the ionizing radiation we absorb every year comes from radon. Medical X-rays account for 11 percent, consumer products for three percent and cosmic rays, eight percent. Some of our organs are affected more by radiation than others, and not everybody is affected in the same way.

One-Third of IFMA Members Test For Radon

According to a survey of environmental factors in the workplace done by the IFMA, one-third of those surveyed now monitor for radon in their buildings. This is up from only eight percent reported in a similar survey done three years earlier. But just what constitutes a "large" dose, and just how much we really need to worry about containing radon, remains debatable. Part of the problem is that to estimate the risks posed by radon, the results from studies involving underground miners exposed to high concentrations of radon decay products must be interpreted as they would apply to an average building. Based on those estimates, the chances of contracting lung cancer are about one in 250, or 0.4 percent—10,000 lung cancer deaths per year in this country. Of course, it is much higher for individuals who spend a large amount of time in homes or buildings with severely elevated radon levels. By contrast, other pollutants we are forced to live with are generally regulated so their chances of causing premature death from exposure are no more than 0.001 percent. Risk from exposure to asbestos is about 0.02 percent, while the risk from breathing in secondhand cigarette smoke is about 0.1 percent, and from driving an automobile about two percent. We are all going to die someday. But we also are willing to take on certain accepted risks—nobody forces us to leave home or drive an automobile. But we do take an interest when we can eliminate some of our risk, and we also tend to be more committed when the exposures are not of our own volition.

Abatement and Testing

So why haven't we heard more about radon? For one thing, little money is available for radon abatement. According to the U.S. EPA, The Indoor Radon Abatement Act of 1988 (Title III of TSCA) establishes a long-term national goal of achieving radon levels inside buildings that are no higher than those found in ambient air outside of buildings. While technological, physical and financial limitations currently preclude attaining this goal, the underlying objective of this document is to move toward achieving the lowest technologically achievable and most cost effective levels of indoor radon in new residential buildings. These methods can be applied successfully in mitigating radon problems in some existing nonresidential buildings. However, their effectiveness when applied during construction of new nonresidential buildings has not yet been fully demonstrated. Therefore, the EPA does not recommend that, pending further research, these building standards and techniques be used at this time as a basis for changing the specific sections of building codes that cover nonresidential construction. The Joint Commission on Accreditation of Healthcare Organizations has no standards addressing radon in hospitals, but an article published in its *Plant, Technology & Safety Management Series* points out that radon testing takes little time and remedial costs, if any, are often low. For instance, said the Commission, "it is unlikely that radon levels would be high throughout a hospital because most rooms are under positive pressure. If there is a problem, most often it is confined to a room or two, and hospitals are generally able to remove this potentially significant risk with little expense." It would be impractical to test all the rooms in a hospital. Instead, try to test:

- All employee rooms in ground contact that are occupied frequently by the same persons.

- Some unoccupied areas such as basement storage rooms where a worst-case reading would be expected.

- Any areas on the second level that are above untested first-level rooms.

- Three or four additional locations on the second level for a spot check, representative sample.

Unfortunately, exact test methods are sometimes in doubt— some critics even claim that short-term testing is without value, or at least the results from them can be too easily skewed to be of much value. Nor are there many quick, buy-this-product solutions available, once testing is completed, that can be easily or simply implemented to remove radon's hazardous effects. Vacating the building or house is one alternative—not a very viable one for most people, especially since they cannot sell the property to somebody else without them facing the same risks. Even tearing down the building and constructing a new one may not help. That is why mandatory radon testing on all real estate transactions is becoming law in many states.

Remediation
Remediation devices for radon can include basement or crawl space ventilation, one of the easier solutions to a complex

problem. Easier, but not necessarily more workable or beneficial. This is because increasing the rate of ventilation leads to higher fuel bills, since more outside air will find its way in to take the place of the air being expelled. This can be offset by using an air-to-air heat exchanger, which can cost $500-$2,000 and an additional $150 or so per year in electricity.

"Our [heat recovery ventilator, or HRV]," claims one manufacturer, "provides fresh indoor air without wasting energy dollars ... The addition of fresh air can help reduce the indoor levels of toxins given off by cigarette smoke, cleansers, sprays, solvents, paints and glues..." And, presumably, radon. Heat is captured in an energy recovery core before being expelled. Without using an HRV or ERV, merely increasing the amount of ventilation also tends to create a negative pressure in the area being ventilated—air being moved out must be replaced by more incoming air, and this can mean more radon gas, from below. It is sort of like using a small pail to bail out a quickly sinking lifeboat. Sealing cracks and holes in the basement walls and floor is another hit-and-miss practice, although it is one worth trying.

The problem is that often these entry points are not easy to detect. Sealants and caulks have to be frequently renewed or at least monitored for breakdown, and many buildings will inevitably continue to leak despite some of the most spirited of sealing efforts.

So unhappily, some radon mitigation systems can have the opposite effect, drawing in even more radon-laden air. Proper roof venting of the exhausted air, not ground level release, is necessary in order to keep the exhausted air from re-entering the building and raising concentrations even more by adding to the newly infiltrating radon-laden air. Excavating to remove any radon-laden soil or rock is probably the most sure-fire solution, but it is also the most expensive and often the most impractical. How deep does one go to ensure results? And how does one remove enough soil from the foundation of an existing building to impact radon levels without risking significant structural damage?

Positive Pressurization

Providing positive pressurization is sometimes a practical and affordable solution. Differential pressure is the difference in air pressure between inside a building and outside. Positive pressures can halt the flow of radon gas into a building, and also keep out other contaminants, such as dust and dirt, and odors from such outdoor activities as hauling away garbage, resurfacing the parking lot or tarring the roof. Positive pressures from within also help to ensure proper venting of combustion products, preventing backdrafts and carbon monoxide leaks. Enconsys Inc. of Des Plaines, IL, supplies an "Air Manager" building pressurization system that measures the amount of makeup air to keep demand to what is needed, preventing wasted energy and overly taxing the mechanical equipment beyond what is required. It should be noted that most buildings will (or at least should) have a slight negative pressurization from release of waste gases such as kitchen exhaust fans, bathrooms fans, smoking lounge fans, combustion equipment, etc. Buildings equipped with fume hoods such as laboratories and manufacturing plants will exhaust even more air. Any contaminants will tend to remain inside the building. In fact, whenever a door or window is opened even more outside air will be pulled in because of this negative pressurization, bypassing any filters.

Separate zones can be set up inside a building to control this flow of air, encouraging it to go in a particular path or direction. One could, for instance, "push" the air from a clean manufacturing operation towards a less critical office area and then through a lobby or warehouse-storage space before it is exhausted to the out-of-doors. Restaurants commonly benefit from keeping a kitchen under slight negative pressure so that cooking smells, steam or heat do not find their way into the dining area.

Radon awareness and remediation varies widely by country and geographic region. One might also conclude that it also varies by economic condition. Poorer countries often have more pressing matters to worry about. Radon is still well down on the list of woes for most regions of the globe that still experience war, famine, disease and contaminated drinking water.

Still, some building owners would do well to know that they are working and living in structures that have hundreds of times the recommended acceptable safe level for radon—even assuming a rather wide range and plenty of leeway in estimating just what constitutes a "safe" level. As we stated previously, the U.S. EPA advises the structure be fixed if the average of two tests is 4 pCi/l or more.

The U.S. EPA calculates that a reading of 200 pCi/l is equivalent to the health risk of smoking four packs of cigarettes per day.

One Out of Five Schools Has Radon

The U.S. EPA estimates that nearly one in five school buildings has a radon level above 4 pCi/l. It recommends that all school buildings be tested, at a cost of $500-$1,500—while mitigation of a problem building or single classroom can range $3,000-$30,000, depending on the scope of the problem and the type of mitigation used.

The EPA Office of Radiation and Indoor Air is supporting an evaluation program to study large buildings for radon. However, much depends on how much of a cost-cutting mood Congress is in over the next few years. The funding for asbestos, for example, was severely cut for fiscal year 1996, with state and local governments expected to pick up the difference. The U.S. EPA, in its booklet *Home Buyer's and Seller's Guide to Radon,* states that "nearly one out of every 15 homes in the U.S. is estimated to have elevated radon levels." It recommends testing for all homes (the same would apply to commercial buildings, but the publication is directed to consumers), either by hiring a certified contractor or by do-it-yourself methods. There are short- and long-term tests: If one uses a short-term test and results indicate a "high" reading, a longer-term test for accuracy should be taken. A long-term test can take 90 days or more. Each state has a list of approved radon testing practitioners who have passed a state testing program.

Passive radon detectors do not require an external power source to operate. These include charcoal canisters, charcoal liquid scintillation devices and alpha-track detectors, often found in hardware stores and home centers. Electret ion chamber detectors are another means, usually available only through laboratories. These tests are all relatively inexpensive. *Active* devices should be used only by trained testers, and require some skilled interpretation of results. In any case, detectors are placed in the lowest level of a building. There are a number of other guidelines to follow, mainly that short-term tests be conducted with windows closed and doors only opened for normal entry and exit purposes. Exhaust fans should not be used during the test period as they can skew the results.

Uncertainty has cloaked radon in a shroud of inaction, or at least delayed action. The book *Indoor Pollutants* states: "The database on sources and source strengths of indoor radon is just beginning to be established. Initial attention focused on building materials and groundwater. Recent evidence from regional studies in the United States points to ground soils (under buildings) as perhaps the major source of radon. Only a small number of buildings in the United States have been measured for radon and radon progeny." Although this book is more than a decade old, progress has been slow and the situation is largely unchanged.

One book critical of radon policy in the United States is *Element of Risk: The Politics of Radon* by Leonard A. Cole, which points out some disparities in federal policies regarding radon over the years, and illustrates some of the difficulties whenever science and politics clash. The author also questions the recommended remediation level of 4 pCi/l, which has always been largely arbitrary. "Although generally out of public view, scientists had begun to debate the extent of the danger, the manner of addressing it, and the proper role of government in the process," said Cole. "Some experts contended that the government was not acting with sufficient vigor to protect the public. Others argued that the gas posed virtually no risk to the general citizenry and that no indoor radon policy was warranted." When faced with a potentially expensive and at least inconvenient radon remediation task, the response of the building owner is not unexpectedly, How great is the risk? How necessary is this action?

"Health House" is an ongoing public health project designed to incorporate a number of products and practices which make homes safer and more healthy to live in. It is a joint effort overseen by the American Lung Association, usually with a local builder and a utility company. Air cleaning devices, upgraded insulation, non-polluting, energy-efficient appliances, low-formaldehyde lumber, etc. are all used to keep the level of contaminants low in this model home. Also commonly used is a passive radon abatement system. This consists of a permeable layer of gravel and/or sand laid down before the building's foundation is constructed, topped with a layer of impermeable, 6 mil polyurethane or similar plastic sheeting. Soil gas should flow through this layer to the exterior of the building, where it is harmlessly vented. This method adds very little to the cost of new building construction. Since buildings cannot be tested for radon before they are enclosed, this "passive system" may never be necessary if radon levels are deemed low enough after construction. However, if a radon test done after the building has been sealed and is ready for occupancy reveals a high level of radon, remediation at this point can be costly and difficult. But with the "Health House," small exhaust fans would be installed in the passive system at this point to make the passive system into an "active" one. The fans would serve to draw the soil gas up through sealed vent stacks in the home, and out the roof just like a chimney. Commercial applications are possible, but have not been tried on any sort of widespread basis so far. Radon is and has been a political football, and further action on it will depend on which way the legislative winds blow—at least, until there is a more definitive national tone on radon and just how much danger we are in from exposing ourselves to it.

The Final Word

Radon may be a silent killer, but it is also invisible in another way: when it is hidden from public concern. Many people do not even know what radon is, much less why it can be dangerous. Still, managing a facility for possible radon problems should be a part of every facilities manager's or building owner's total environmental awareness program. Effective means are at hand for detecting high levels of radon, as well as ways to mitigate its hazards to building occupants.

ENVIRONMENTAL TOBACCO SMOKE (ETS)

One of the great enemies of indoor air is environmental tobacco smoke (ETS). This comes as no surprise, as it has been decades now since researchers first identified the many drawbacks of smoking cigarettes. It was only relatively recently that research confirmed that secondhand smoke, that which is exhaled by smokers, or environmental tobacco smoke, which is present whenever tobacco is burned regardless of whether or not you have a cigarette in hand, is a health hazard for nonsmokers. This announcement came only after much research, and despite a movement as early as 1971 to ban smoking in public places. Studies then were inconclusive, however, because of the complexities of the various substances found in cigarette smoke. A report by the National Research Council in 1981 shows the prevailing mood and lack of clear evidence at the time: "Information on passive exposure of nonsmokers to tobacco smoke has never been systematically obtained, except in a limited number of epidemiologic investigations." Although elsewhere in that study we see: "Children whose parents smoke have been shown in some studies to be more likely to have respiratory symptoms, bronchitis and pneumonia as infants."

The study also stated, "Tobacco smoke has shown some evidence of being a major contaminant in many indoor environments," and, "Tobacco smoking indoors can contribute to or cause increased concentrations of respirable particles, nicotine, carbon monoxide, acrolein, and many other substances in the smoke."

After secondhand cigarette smoke was classified a group A carcinogen—one of only 15 such pollutants, including asbestos, radon and benzene—the U.S. EPA was unequivocal in its report, "Setting the Record Straight: Secondhand Smoke is a Preventable Health Risk." Every year, according to the report, an estimated 150,000-300,000 children under 18 months of age get pneumonia or bronchitis from breathing secondhand tobacco smoke. Secondhand smoke is a risk factor for the development of asthma in children and worsens the condition of up to one million asthmatic children. The U.S. EPA "absolutely stands by its scientific and well-documented report ... Virtually every one of the arguments about lung cancer advanced by the tobacco industry and its consultants was addressed by the [Science Advisory Board]. The panel concurred in the methodology and unanimously endorsed the conclusions of the final report."

The report was endorsed by the U.S. Department of Health and Human Services, National Cancer Institute, Surgeon General and many major health organizations. According to the U.S. EPA, only secondhand smoke among class A carcinogenic pollutants has been shown to cause cancer at typical environmental levels.

The U.S. EPA estimates that 3,000 nonsmokers die each year in this country from lung cancer caused by secondhand cigarette smoke.

Tobacco Epidemic?

According to "The Global Tobacco Epidemic," an article in *Scientific American*, in the mid-90s, "Tobacco also drains society economically. The University of California and the Centers for Disease Control and Prevention (CDC) have calculated that the total health care cost to society of smoking-related diseases in 1993 was at least $50 billion, or $2.06 per pack of cigarettes— about the actual price of a pack in the U.S. ... smokers take from society much more than they pay in tobacco taxes." Cigarette marketing expenditures are up considerably from 1975-1988, and while domestic consumption of cigarettes is down somewhat, total production of cigarettes is little changed from where it was a decade ago, helped in part by higher exports. This is despite the fact that the United States has one of the lower rates of taxation on tobacco versus such countries as Norway, Denmark, the United Kingdom, Sweden, Ireland, Finland, Germany, Canada, Australia and France. In the U.S., tobacco marketers spent a remarkable $31.6 million on outdoor advertising in the first quarter of 1995, making it the highest spender by far. This was partly because the major tobacco companies were forced to scramble to maintain share in a dwindling market, but it is a substantial market nonetheless.

A U.S. Food and Drug Administration proposal included:

- A tobacco vending machine ban.

- No sales of tobacco products to anyone under 18.

- A ban on tobacco outdoor boards within 1,000 ft. of a school and limiting them to text or black-and-white images only, and text-only ads in magazines with greater than 15 percent readership by those under 18.

- Restrictions on brand sponsorship of events by tobacco companies. These could be sponsored in name only, without use of corporate mascots or images, with the ban extending to the color of cars in racing events.

- A ban on most contests and giveaways.

- Text-only ads used in direct mail.

- $150 million in annual media anti-smoking messages, paid for by tobacco advertisers.

Smoking Is On Its Way Out

The number of smokers in the United States is way down, from a peak of 42 percent around 1965 to about 25 percent in 1990. However, this trend downward has leveled off over the past 10 years and was stagnant from 1990-1992. In fact, tobacco use is growing slightly among teenagers. Fifteen percent of white males in grades 9-12 reported being frequent cigarette smokers in 1995; for females, it was almost 16 percent.

Down through the centuries, tobacco has been both lauded and loathed, occasionally condemned by kings and prime ministers, but always surviving into what must have been truly the golden age of tobacco—America in the 1950s, when mass advertising campaigns on television reached millions of viewers with memorable lines like "Lucky Strike Means Fine Tobacco" and images of the Marlboro Man. The Philip Morris Company sponsored *I Love Lucy* in the earliest and one of the most widely watched television sitcoms of all time. Smoking and America went

Figure 2-1. Cigarette consumption per adult in the United States. Courtesy: U.S. Department of Health And Human Services.

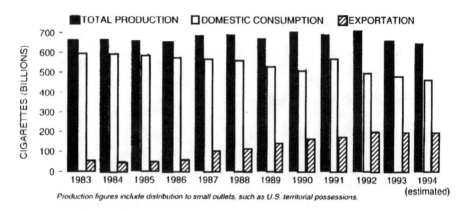

Figure 2-2. U.S. cigarette distribution. Courtesy: *Scientific American.*

hand in hand, it seemed, at least until tobacco ads were banned from television in 1970.

Per capita consumption continued to soar until its peak in 1977, despite the first widespread official recognition of the dangers of cigarettes released by the U.S. Surgeon General in 1964. Tobacco ads in 1970 represented about seven percent of total ad revenue for television.

It was no surprise that beginning in the 1980s, scientific and medical research began to focus on the effects of tobacco smoke on non-smokers. According to a paper written by Jonathan M. Samet for *Indoor Air and Human Health,* "In summary, at present, only nine published investigations provide data directly relevant to the hypothesis that passive smoking is a risk factor for lung cancer ... New approaches for studying passive smoking and lung cancer are clearly needed."

The overwhelming avalanche of evidence and condemnation by health and environmental groups did not come until the 1990s. Oddly, while cigarette smoking has declined in the U.S., it has gone up on a global scale, with countries such as China increasing their per capita consumption as much as 11 percent each year. It is no surprise that the U.S. had a $4.2 billion trade surplus in 1990 from its tobacco exports—accounting for more than a third of our total national agricultural export surplus. Although, in Canada a tougher stance on smoking includes rather stark warning labels on cigarette packs: "Smoking can kill you." One study showed that 83 percent of the schoolchildren shown the warning label remembered it, versus about six percent for the softer "Surgeon General's warning" in the U.S.

R.J. Reynolds Tobacco Company began testing low-smoke cigarettes—based on technology that *heated* tobacco rather than burning it. This product was especially aimed at reducing second-hand smoke. But at the same time antismoking efforts from other sources stiffened, including additional taxes on cigarettes, smoking bans in many buildings and an attack at the federal level on tobacco subsidies in agriculture. Still, Philip Morris CEO Geoffrey Bible said on the delivery of strong second-quarter earnings in 1995: "We continue to grow rapidly, led by our extraordinarily successful worldwide tobacco business."

Research was slow in coming because the impact of ETS is hard to measure versus that of some other environmental contaminants. But an increasing amount of evidence showed that having no-smoking areas in the workplace was an easy way to guarantee that nonsmokers would be free from the effects of tobacco smoke. These findings put additional burden on building owners and facility managers to segregate smokers from non-smokers: the more easily followed route was to ban smoking from the workplace completely.

In 1986, only three percent of employed adults reported a total ban on smoking at work. According to an article that appeared in the *Journal of the American Medical Association (JAMA)*, research indicates that the only way to protect nonsmokers' health is with a smoke-free work site.

ASHRAE Deals A Potentially Crippling Blow

What may have been a crippling blow to allowing smoking in buildings came with development of ASHRAE's new replacement standard for its 62-2001. According to this standard, it would be virtually impossible to allow smoking in a building and still be able to ensure quality indoor air—in other words, to be in compliance with the standard.

As the time this book was being written, this standard was still not final. In addition, many buildings may not be able to meet this rather lengthy and somewhat optimistic ideal. However, committee members said they could not dwell too much on enforcement of the standard, or whether most buildings or building owners would be able to comply with it. The real goal of the standard is a research and technical one, not one of implementation or enforcement. It is a road map, not a national law. And most committee members agreed: smoking and indoor air quality simply do not mix.

In healthcare facilities, the situation is even more significant. The Environment of Care standards by the Joint Commission on Accreditation of Healthcare Organizations generally require that hospitals and mental health organizations maintain and enforce no-smoking policies. However, some flexibility remains for patients receiving long-term care, including:

- Chronically mentally ill patients.
- Long-term or intermediate care and skilled nursing patients.
- Forensic psychiatry patients.
- Post-acute head trauma (social rehabilitation) patients.

Some health care organizations have gone so far as to screen job applicants for smoking status, giving preference to nonsmokers. If no-smoking policies are in effect, it is a good idea to arrange an outdoor designated smoking area which is also out of inclement weather, equipped with benches, shade, etc.

No-Smoking Is No Guarantee

In most smoking lounges, smoke-filled air finds its way into the building ventilation system. An ASHRAE paper authored by engineers from R.J. Reynolds told of the study of a Washington, DC, bar and restaurant which had only one ventilation system for both its smoking and nonsmoking sections. Even worse, this street-level portion of a high-rise building was operating with zero outdoor air. Space was a problem. There was little room to add additional ductwork or filtration. A simple solution was to add a fan to exhaust some of the smoke-filled air and introduce fresh air from the outside.

To have a proper smoking lounge, the room must have a sufficient supply of makeup air (not all of it need come from

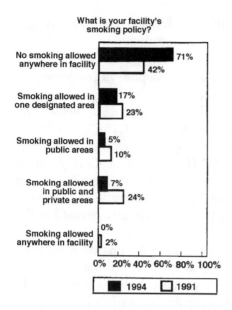

What is your facility's smoking policy?

No smoking allowed anywhere in facility: 71% (1994), 42% (1991)

Smoking allowed in one designated area: 17% (1994), 23% (1991)

Smoking allowed in public areas: 5% (1994), 10% (1991)

Smoking allowed in public and private areas: 7% (1994), 24% (1991)

Smoking allowed anywhere in facility: 0% (1994), 2% (1991)

■ 1994 □ 1991

Figure 2-3. 1994 IFMA IAQ survey. Courtesy: IFMA.

outdoors) and exhaust must be great enough to create negative pressurization. Negative pressures keep the smoke-filled air from spilling over into surrounding areas. ASHRAE recommends 60 cfm for each occupant. A smoke detector tied to an exhaust fan to control the air pressure in the space is the best solution if a smoking lounge is to be permitted. If the space is unoccupied, the exhaust fan could be turned off and would not waste energy by moving out conditioned air. This sort of a solution can prevent an "us versus them" showdown among employees, or smokers versus non-smokers, protecting the rights of both groups.

R.J. Reynolds has conducted many studies and offered suggestions for improving the air quality in restaurants and other buildings—the company obviously has a vested interest in this respect. R.J. Reynolds conducted a study of restaurants in 1993 in Denver and Washington, DC. A test and balance firm was used to measure supply air, return air, pressure drop across filters, etc. on all HVAC units. The same information was collected on the makeup air unit and exhaust fans in the kitchen. Supply air flow rates for each supply air diffuser were recorded.

According to the R.J. Reynolds' engineers, Stephen C. Curl, PE, and Hoy R. Bohanon, PE, the purpose of this testing was to:

• Measure indicators of the movement of air inside the restaurant.

• Determine mechanically supplied outside air ventilation rate.

• Determine pressure relationships relative to the inside and outside.

- Provide an assessment of the mechanical condition of the HVAC equipment.

- Create a baseline against which to compare improvements.

The paper, which was delivered at a conference sponsored by the National Coalition on Indoor Air Quality (NCIAQ), concluded that improvements to the ventilation systems in the restaurants greatly improved thermal comfort. "Thoughtful location of smoking and nonsmoking sections, filtration and strategic balancing can improve air separation of smoking and nonsmoking sections," concluded Curl and Bohanon. "A test and balance appraisal is the first and the last step in improving restaurant ventilation."

Still, the movement to ban tobacco smoke has been steadily picking up momentum. According to a survey done by IFMA, compared to a similar survey three years before, significantly more facilities do not allow smoking in any area of their buildings, up from 42 percent to 71 percent.

Meanwhile, BOMA made its own point clear: "We don't want to accommodate smokers, we want to ban [smoking]." Acknowledging smoking in the workplace not only is a detriment to indoor air quality, according to BOMA, but results in lost productivity—why pay workers to smoke? If smoking is limited to certain areas, perhaps even out-of-doors, these smokers must presumably be accommodated with free time to get to these smoking areas. On the other hand, some people are more accommodating. If workers are going to smoke, is it not just to give them a place to do so, without having to leave the building to stand out in the cold or rain? And would that not just lead to even more lost productivity?

Not all share that view. Paul Cammer, president of the Business Council on Indoor Air, said he is not as "paranoid" when it comes to ETS in the workplace, saying, "There are thresholds and ways to deal with it." BCIA "was established to ensure that legislative and regulatory responses to indoor air quality concerns are based on information that is both accurate and comprehensive—before government policy is set." Cammer took issue with a BOMA statement that smoking bans are cost-free to institute. Employees will go outside if their buildings do not have a designated smoking area, he said, taking away from time on the job. On a conference panel on the merits and demerits of indoor tobacco smoke (Indoor Environment in Baltimore) Brooklyn attorney Mark Diamond even asks if employers must *by law* accommodate tobacco smokers. If smoking is indeed an addiction, he reasons, should not employers have to recognize this and make accommodations for them under the Americans with Disabilities Act (ADA), just as you would with an employee who is confined to a wheelchair, or who is blind? Said Diamond: "Owners and managers of public buildings will have to re-integrate smokers into the general population or provide separate-but-equal accommodations that are at least as comfortable and accessible as non-smoking areas." This raises some interesting questions, and ones that it may take years to sort out the answers.

Smoking, in fact, always causes spirited debate. At ASHRAE's Summer Meeting in San Diego, *Air Conditioning, Heating and Refrigeration News'* editor-at-large Thomas Mahoney said one such discussion "degenerated into a food-fight version of the TV talk show *Crossfire*, where debate quickly turns into shouting." Title of the talk, which featured a panel which included a tobacco industry lobbyist, was "IAQ and the GOP: Indoor Air Quality and the Newt Congress."

Smoking was and continues to be a hot political football.

Smoking Banned In Restaurants

Early in 1995, New York City adopted a ban in most of the city's restaurants. Smoking is allowed only in restaurants that seat fewer than 35 people, in separate bars of larger restaurants, or in stand-alone bars not part of a larger building.

"It's a disaster," said one restaurant owner, who said business was down because of it. Petitions were being circulated to get the city's lawmakers to reconsider. One restaurant in Greenwich Village reportedly built a special smoking section, with separate AC, ventilation and air quality monitors, at a cost of $30,000 just for one of its favored customers—actor Jack Nicholson. Nicholson is an avid cigar smoker. Another phenomenon—an unusual one— saw a sudden trend toward special gourmet meals prepared for and served on behalf of cigar smokers like Nicholson, David Letterman, Lee Iacocca and George Burns. Many of these—fortunately for those non-smoking patrons—were served on outdoor patios or when the restaurants were closed to regular customers.

Maryland, California, Utah, Washington and Vermont all have enacted a smoking ban in workplaces except for bars and restaurants—of course, there is a chance that others have followed, or that some of these have been repealed or altered, in view of the volatility of this issue. In Maryland, such a ban on tobacco would have included even taverns and restaurants were it not for concerted last-minute lobbying efforts right up until the vote was scheduled. In Vermont, an earlier workplace ban on smoking now includes restaurants. About 200 individual municipalities also ban smoking in the workplace. For updated information call Action on Smoking and Health (ASH), (202) 659-4310.

Also in the headlines: The Dunkin' Donuts chain announced a ban on smoking in its 3,000 restaurants by month's end. The local doughnut shop would seem like one of the last bastions of escapism, where coffee and a cigarette go hand in hand. Can a broader ban be far behind? This was despite a comment by one outlet co-owner, who said that 70 percent of his customers are smokers. But one customer may have echoed the sentiments of many: I've resigned myself to the no-smoking rule, he said. I'll just get my coffee to go.

The Final Word

Tobacco smoke is one old-time problem in buildings that has recently been recognized as being more than just a nuisance or distraction. It is interesting to note how far the battle has gone in the course of just a few years. It was not very long ago that smoking cigarettes was part of the average office work experience, as welcome as the water cooler or coffee pot. It was the same everywhere, welcome in our restaurants, advertisements, a part of everyday life. Today, lines are being drawn between smokers and nonsmokers everywhere, and nowhere is it more apparent than in the workplace. Do we protect the rights of smokers, or only the right of non-smokers to be segregated from smokers? For facility managers, it is an easy decision to ban smoking and thus avoid any possible problems. This may or may not be an option, for philosophical or other, more practical reasons. For those who continue to allow smoking, the task of accommodating it can be difficult, although there are ways to try to improve the situation and thus serve the needs of everyone. Restaurants and other public facilities should especially note this. However, some argue that "no smoking" is the only reasonable smoking policy. The debate is far from over.

LEGIONELLOSIS AND RELATED ILLNESSES

Feeling uncomfortable or sleepy, not being able to perform the job as well as one might, or having unhappy-unsatisfied occupants are all problems that can relate to HVAC systems and building IAQ. They are problems the facility manager must come up against in day-to-day operations. But much more serious, deadly in fact, is the potential problem of *Legionella* found in some unsuspected nooks and crannies as innocent as public drinking fountains. Is this one of the new major plagues of the 20th century? Certainly, its risks can be minimized with proper attention to the building's HVAC and plumbing systems.

Legionellosis is an illness that is occasionally of serious concern in the HVAC industry. One form of it is Legionnaires' disease, first noted after an outbreak in 1976 among an American Legion convention held in Philadelphia, eventually killing 29 people. It is a disease that only affects 5-15 percent of those exposed to it, but it is fatal in 15 percent of those cases. An estimated 25,000 people annually die from the disease. Initial symptoms include general physical discomfort (malaise) and muscle pain (myalgia) followed by fever, chest pain, shortness of breath, vomiting, headache, weight loss, diarrhea, coughing, recurrent chills and abdominal pain. This can lead to kidney impairment and loss of protein in the urine, and pneumonia. All surface water supplies are possible breeding grounds for *Legionella*. Chlorination does not always kill the virus, since *Legionella* grow inside large protozoa, and slime layers can protect them unless the chlorine level is high enough and left for a specified amount of time until flushed from the system. Mechanical cleaning devices to break up the slime help, but carry no guarantee.

Other noteworthy outbreaks of Legionnaires' disease occurred in 1979, 1980, 1984, 1989, 1991, 1995, and 2003, some of these with fatalities, in the United States, with other outbreaks reported in England, New Zealand and Australia. In the summer of 1977, it struck a new hospital located in one of the wealthiest parts of Los Angeles, the Wadsworth Medical Center, killing 16 over a period of a year. In July 1995, health officials searched for the cause of an outbreak that killed an 83-year-old woman in Chambersburg, PA, then two more, hospitalizing at least a dozen. It was finally traced to a hospital, which reacted with a three-day bleeding of its cooling and hot-water systems. Other cases were discovered in August of that year in Minnesota.

Ten people contracted the disease in 1991 in a Social Security Building in Richmond, Calif. Two died. A subsequent lawsuit filed by Aetna Life & Casualty Insurance Company argued that since the disease can be prevented easily enough, is there evidence for negligence when these outbreaks do occur? The suit was settled out of court, but repercussions are thought-provoking.

Some authorities say Legionnaires' Disease may be more prevalent than we think, because many cases go unreported. Some experts estimate outbreaks of up to 25,000 cases per year.

Lower Water Temperatures

It is one of the great paradoxes of our modern litigious society that lower hot water temperatures in water tanks can lead to safer havens for harboring *Legionella*. Temperatures are occasionally lowered by facility managers to save money by reducing energy costs, but more often to avoid lawsuits from scalding. Just as McDonald's had to lower the temperature of the coffee it serves after a woman burned herself with a cupful, many water heaters are having their operating temperatures turned down to avoid accidental skin burns in restaurants, hotels, etc. ASHRAE Standard 100P, "Energy Conservation in Existing Buildings," in addition to a number of energy-saving recommendations to the HVAC system, lighting and building envelope, stated: "Domestic hot water shall not be hotter than 120°F (49°C) measured at the closest tap to the water heater." An exception was made for systems dedicated to heating water for sterilization purposes and for a system utilizing a water heater to meet both domestic hot water and space heating requirements. It is better, from a microbe-killing viewpoint, to keep this stored water hot, then to temper it on its way to its final use.

Pontiac Fever

Another form of the disease, Pontiac fever, does not cause pneumonia but a flu-like illness which is usually over within 48-72 hours. It often goes undiagnosed and therefore untreated, except to treat its symptoms just as one would any other cold or flu. Similarly, many cases of Legionnaires' disease go unreported because doctors are busy treating it as pneumonia, without regard to the source.

Pontiac fever was identified and named for an outbreak in Pontiac, MI in 1968. Fever, chills, headache, muscle ache, nausea, diarrhea and sore throat generally accompany it. It will infect a far wider percentage of those who come into contact with it than Legionnaires' Disease. It was first traced to the stagnant water in a drip pan from an evaporative condenser in the building's basement.

Humidifier Fever

Humidifier fever is traced to office humidifiers, often afflicting susceptible employees shortly after they show up for work after a weekend off. Flu-like symptoms, muscle pain, fever and lack of energy follow. A chest X-ray will rule out pneumonia, although the lungs may be impaired and sound congested.

> **An estimated 25,000 people die each year from Legionnaires' Disease.**

It is estimated there are 50,000-100,000 cases of Legionellosis annually, along with 2,000-6,000 deaths (according to the U.S. Centers for Disease Control) but most of these go unreported because the disease is not widely known or easy to diagnose. Many cases are lumped together and treated as pneumonia of unknown origin.

It was once thought the prime source of the organisms was in cooling towers or evaporative condensers; but more recently, potable water found in faucets, hot tubs, fountains, whirlpools and water heaters have been found to be the source. "The major mode of transmission," according to Victor Yu, MD, writing in an ASHRAE paper, "from water to human is likely aspiration (water within the mouth bypassing host defenses and entering into the lungs. Aerosolization can occur from nebulizers and humidifiers. Outbreaks of Legionnaires' disease linked to cooling towers have all but disappeared."

Actually, the *Legionella* organism is not uncommon, and can be found in most lakes and streams (pond scum will do fine), especially warm ones, but it does not generally result in illness. It has to be present in sufficient number; the victim's immune system must be lowered or susceptible to the disease; and, this being important from an HVAC point of view, it must be somehow transported (aerosolized) for transport to the host's lungs. It is not passed on from person to person. Nor can one get it from drinking water. It is difficult to see with a microscope unless treated with silver, and is resistant to many antibiotics. Larger particles will generally settle in the upper and lower respiratory tract, where they are expelled, rather than reaching the lungs.

Some thought they knew where to place the blame. "In the case of *Legionella*," wrote Laurie Garrett in her book *The Coming Plague: Newly Emerging Diseases in a World Out of Balance*, "a new human disease had emerged in 1976, brought from ancient obscurity by the modern invention of air conditioning." The 1950s, this book reminds us, dawned a wondrous, confident era when technology was handing us the answers to a number of old problems, including eliminating through vaccines and better living conditions many of the age-old scourges that once decimated our population. Scientists were optimistically predicting the permanent eradication of many common maladies—polio, smallpox, tuberculosis, etc. Guess what? Disease in its latest, most dangerous form is back—back with a vengeance. Not only did we fail to eliminate some of these forever from the face of the Earth, but their replacement brethren abound, looking for new, unsuspecting victims to latch onto. Ours is truly a shrinking world. Little-known diseases like the deadly Ebola virus in African nations are finding their way into export products including live ones such as monkeys used for research purposes, onto airplanes bound overseas, encouraged by the clearcutting of rainforests and free trade agreements. Hospitals are trying to cope with tougher, drug-resistant strains of tuberculosis. The author also noted that many other "new" diseases may be lurking, waiting for the right delivery system or triggering mechanism: "Chagrined by the events of 1976, the U.S. public health community looked to the future, for the first time in the late twentieth century, with a vague sense of unease," wrote Garrett.

To help in the battle against *Legionella*, ASHRAE suggests that air intake ducts should be closed during cleaning if they are within 30 meters (100 ft.) of a cooling tower being decontaminated. Respirators should also be worn by workers during this cleaning process.

Guidance is rather skimpy, and there is movement afoot to recommend and/or require more preventive measures.

Maintenance the Key

Preventative maintenance is essential to reducing the risk from *Legionella*. Not just regular maintenance either, since this is at best a waiting-for-something-to-happen policy which attacks the disease only after an outbreak is reported. Since 1976, updated standards for treating cooling towers and air conditioning systems have grown more stringent because of *Legionella*. Treatments added to cooling towers to prevent corrosion do not control *Legionella*. Bio treatments help, but do not kill directly, as noted earlier— they only keep it from growing and prospering by killing the nutrients it needs. Chlorine is recommended to kill the organism directly. If chlorine is used, consult the manufacturer of the cooling tower first, as chlorine can be extremely corrosive or damaging to certain components.

Several factors affect the growth and reproduction rates of *Legionella*. A temperature range of 93.2°-100.4°F is optimal. However, some species can survive and grow nicely outside of this range. *Legionella* has only two other nutritional requirements: iron and the amino acid cysteine.

Advocating a proactive maintenance program for the control of *Legionella*, Mark Hodgson, Clayton Environmental, Edison, NJ, notes that current health practices in the United Kingdom and those proposed for the pending OSHA workplace IAQ rules in the United States should minimize risk.

Hodgson recommends a four-part approach to the problem:

1.   Conduct a review of *Legionella* sources and associated health

hazards. Where a significant health risk is identified, conduct a risk assessment that considers the potential for amplification and dissemination of infectious agents as well as the susceptibility of the population that may be exposed to any bioaerosols.

2.   Develop a corrective action and preventative maintenance program to control or prevent the growth and dispersion of microorganisms. Provide detailed specifications for minimum performance criteria for critical operations.

3.   Designate a responsible person, develop a record-keeping system, and develop a staff training program.

4.   Conduct periodic audits to monitor performance and to update the program as necessary.

Control of *Legionella* and other microorganisms directly affects the operating efficiency of air and water handling systems. Uncontrolled corrosion can provide a ready source of iron, while a buildup of scale and mineral deposits can form friendly breeding grounds. Properly managed treatment programs can not only serve to identify and correct any problems with the operating systems, but will also reduce health risks and help control any maintenance problems that could be expensive if left untended.

Do Not Overlook Other Sources
While some outbreaks of *Legionella* are traced to cooling towers or air handlers, the domestic water supply should not be overlooked as a possible source of the disease. Frank Rosa writes in his book, *Legionnaires' Disease Prevention and Control*, "With the advent of energy conservation, many retired or disabled homeowners on fixed incomes lower the thermostat setting on their hot water heaters, hoping to lower energy bills. Many elderly people die from what is considered to be pneumonia, so it seems plausible that these deaths could actually be caused by *Legionella* lurking within the hot water tanks of their homes." A sobering thought indeed.

To Test or Not to Test?

Rosa and others take exception to some policies which advocate against testing for *Legionella*. "The reluctance of those in responsible positions to test their systems for *Legionella* seems unbelievable ... however, it is *only through testing* (italics added for emphasis) that one can determine the level of risk and strive to minimize this risk using state of the art technology." The Centers for Disease Control and Prevention in Atlanta has routinely cautioned against testing for *Legionella*, saying tests were unreliable and the dosage difficult to pinpoint. However, some experts now say that routine testing can be valuable and the possible benefits outweigh any arguments against testing.

John Donald Millar, MD, Occupational Health Consultant, PathCon Laboratories, suggests, "Such testing would provide, at a minimum, a useful indicator of how well prevention and maintenance programs are working. Also, if a building owner is held accountable for even one case of legionellosis among employees or visitors, the costs of medical care and workers' compensation—not to mention litigation and liability—would likely dwarf the costs of any reasonable pattern of testing 'high risk' sources."

CARBON MONOXIDE

Carbon dioxide (CO_2) is generally not considered dangerous except in huge concentrations. This is not the case for carbon monoxide (CO). While CO_2 is a normal by-product of human respiration, CO is produced by combustion and is the leading cause of poisoning deaths in the United States. Seldom a winter goes by that there is not a report of a family killed by CO emitted by a faulty furnace or water heater. Such tragedies continue to occur despite advances in technology, despite new equipment, licensing and permit requirements. Even slight exposure can cause flu-like illness. When it enters the lungs, it reduces the ability of the hemoglobin to carry oxygen, leading to asphyxiation. It flows easily into room air, and does not necessarily travel only upward like hot air or smoke. The ideal safe level of CO is zero. Outdoor concentrations generally range from 0-9 parts per million (ppm), although they can peak at high concentrations of around 30 ppm.

Indoor concentrations should not exceed 15 ppm, even though a short term exposure of 100 ppm is not harmful to most adults.

A major manufacturer of water heaters was successfully sued recently when an elderly woman was killed by a gas water heater installed in her mobile home by her grandson. The grandson claimed he was unaware that such appliances must be vented. This kind of poisoning does not take place in residences only, and is in fact all too common.

One facilities manager tells of a CO alarm that went off in a commercial combination laboratory and office facility. Remotely monitored, the alarm was answered by a service technician who quickly phoned the facility—he was happy to discover that somebody was able to answer his call.

It turns out the laboratory, which operates 24 hours a day, had its doors left open to the office portion of the building by the guards that were on duty that night. Exhaust fans were quickly turned on, the doors closed, and no harm was done.

Backdraft

A new danger has arisen since the tightening of our buildings. CO poisoning comes not only from faulty appliances, but because our new super-efficient ones no longer exhaust enough of the air that they burn. CO-laden combustion products are not being completely pushed up our chimneys and vent stacks because the more efficient furnaces consume fuel more efficiently and expel less air. This causes a condition known as backdraft, which means the outside air pressure is enough to keep the inside air from being expelled.

CO detectors have come into great demand just for this reason. There are different types and models, but are generally divided into two types. One type is a paper-like indicator that is attached to a wall and must be "read" by someone to determine presence of CO. A report published by *Consumers Digest* magazine gave a thumbs-down to this type of indicator. Too often, if there is a CO problem it does not help to have an indicator that must be "read." The unfortunate victim might never make it. For residential structures, such an indicator would have to be read constantly, in case there was CO leakage. More practical are the "alert" type devices that signal an audible warning similar to that of a smoke detector. Indeed, the market for these type devices has boomed, with shortages of this device turning up. Better yet, especially for the commercial building owner, is the monitor that will electrically signal a remote sensing station and sound an alarm for an emergency service call.

MULTIPLE CHEMICAL SENSITIVITY (MCS)

Some people may be more susceptible to environment-related ailments, adding to the difficulty of accurately diagnosing the causes or zeroing in on their IAQ complaints. Some people may suffer from what is known as multiple chemical sensitivity—symptoms believed to be triggered by low level exposure to certain chemicals, often after a previous, higher dose or incident has "sensitized" their bodies. Health authorities disagree over whether or not MCS even exists, much less the true dimensions of the problem. Objective, quantifiable indicators are lacking, although some much-needed research was launched in 1992 after a protocol to perform an evaluation of MCS sufferers was established at a National Academy of Sciences workshop. Do not expect results on this debate any time soon, but be aware that MCS was found to be compensable in court (see Chapter 8 for a case study).

Sources
ASHRAE Legionellosis position paper, Atlanta: ASHRAE.

Bartecchi, Carl E., MacKenzie, Thomas D. and Shrier, Robert W., *Scientific American*, May 1995, p. 44.

Curl, Stephen C., PE and Bohanon, Hoy R., PE, "Practical Techniques for Improving Quality in Restaurants." *Environmental Issues in The Workplace II.* Houston: International Facility Management Association, 1994.

Curl, Stephen C., PE, "Effects of Ventilation and Separation on Environmental Tobacco Smoke Concentration," ASHRAE paper, IAQ '95.

Freije, Mathew R. *Legionellae Control in Health Care Facilities*, HC Information Resources, Inc., 1996.

Garrett, Laurie. *The Coming Plague: Newly Emerging Diseases in a World Out of Balance*, Farrar, Straus And Giroux, 1994.

Hodgson, Mark and Thomas, William M., Ph.D., "Control of Health Risks from Legionellosis: A Proactive Approach," *Technical Newsletter*, Clayton Environmental Inc., May 1995.

Indoor Pollutants Report, National Research Council, National Academy Press, 1981.

Invironment Professional Newsletter, Itasca, IL: Chelsea Group Ltd.

Mader, Robert P., "Cross Connection on Ice Machine Sickens Conventioneers in Ohio," *Contractor*, December 1995.

Millar, John Donald, MD, "Legionellosis: To Test or Not to Test?," PathCon Laboratories, *Invironment Professional*, July 1995.

"Philip Morris Smoking," *USA Today*, July 19, 1995.

"Protection from Environmental Tobacco Smoke in California," Journal of The American Medical Association (JAMA), August 12, 1992.

Rosa, Frank, *Legionnaires' Disease Prevention And Control*, Business News Publishing Co., 1993.

Samet, Jonathan M., "Relationship Between Passive Exposure to Cigarette Smoke And Cancer," Proceedings of The Seventh Life Sciences Symposium, Lewis Publishers, 1984.

Yu, Victor L., MD, "The Reservoir for Legionella: Potable Water or Cooling Towers?" ASHRAE paper, 1994.

Chapter 3

HVAC Basics and Indoor Air Quality

To have a working knowledge of indoor air quality, we must also have a working knowledge of building HVAC. While not all IAQ problems are directly related to HVAC, a good deal of them are. An understanding of mechanical systems is essential to maintaining good ventilation and healthy indoor air.

THE HVAC SYSTEM

It was not long ago that we depended on something as mundane as a coal stove for warming our surroundings, enduring excruciating heat in the summer as just another by-product of doing business. Air conditioning has only been around for half a century or so.

Willis Carrier, "the Chief," was born in western New York state in 1876. He was one of those early inventor-pioneers such as Thomas Edison and Henry Ford who had the uncanny ability to combine an interest in pure science with a natural mechanical aptitude. After testing and experimenting, Carrier designed what is said to be the world's first air conditioning system in 1902. Almost immediately, he began to refine and improve on this system. He created an "apparatus for treating air" for the textile industry, then went on to an air cooling and dehumidifying system for a pharmaceutical company. Carrier Air Conditioning Company of America formed in 1908. The Age of Air Conditioning had begun.

Today's modern heating, ventilation and air conditioning (HVAC) system comprises all components working together to introduce, distribute and condition air in a building for human comfort. Components include furnaces or boilers, water chillers, cooling towers, air handlers, ducts, filters, exhaust fans, supply fans, controls, heating and cooling coils and other components that all work together in a unified design to move and condition air. The mission of the HVAC system is to provide thermal comfort, circulate fresh air and dilute contaminants.

Note that not all HVAC systems perform the full range of functions of heating, cooling, ventilation and humidity control. The use of the building, the local climate and other factors come into play. An important factor is age of the design and the prevalent building codes in existence at the time of construction. Consider that one-third of our country's nonresidential buildings, about 1.6 million, were built before 1960. The mechanical systems in them run the gamut from newly installed to the originals to modified originals, some 30 or more years old and running on outdated technology.

To launch an effective IAQ program, therefore, facility managers should conduct a thorough study of their HVAC systems. Because not all HVAC systems are created equal.

HVAC SYSTEM OPERATION

Here is a simplified explanation of how a basic HVAC system works, again noting that specific types and designs will vary in how they handle the basic functions:

Based on monitored interior temperatures or some other control strategy or at a steady rate, the HVAC system allows outside (supply) air to enter the outdoor air damper through an intake.

This air is filtered to remove contaminants, then pulled by a fan into an air handler. The air handler heats or cools the air, then dehumidifies it. The amount of heating, cooling and dehumidifying is based on temperature gauges and humidistats (and is only as capable as the system's design capacity).

Supply air diffusers blow the air into the interior space.

In another part of the space, a return air grille accepts air back out of the room, where some is exhausted outside through a vent (exhaust air), while some enters an air-mixing plenum (return air). In this chamber, fresh supply air is mixed with the return air and the combined air enters the filter and the rest of the loop.

The process takes place at a central HVAC plant and/or a series of stand-alone units each responsible for one or more zones.

Special rooms with high levels of odors or indoor pollution may receive air from the ductwork, but the return air is exhausted via a fan directly to the outside to prevent it from recirculating to other interior spaces.

HVAC SYSTEM TYPES

HVAC systems can be classified in different ways. All-air, air-and-water and all-water systems are the primary choices. These may be centralized or decentralized. They can then be classified as constant air volume or variable air volume depending on how they distribute air. Other classifications describe specific-use systems.

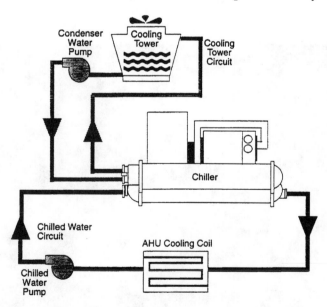

Figure 3-1. A typical water-cooled chiller system. Courtesy: U.S. EPA, Energy Star Buildings Program.

Note that traditional criteria for system and product selection is expanding to include "Green Building" objectives such as refrigerant selection, improved IAQ, energy efficiency, lower sound levels, improved air distribution and smaller product size.

Some mechanical engineers and IAQ investigators believe that specifying future HVAC systems will depend at least in part on the type of materials and furnishings used in the building—in other words, more ventilation for materials that emit more gases—something that would have been considered science fiction just a few years ago. But a higher off-gassing type of carpeting could conceivably one day require installation of a larger-capacity air handler, larger ductwork and so on. On the other hand, trade-offs could conceivably be made: Substituting a lower-VOC type of paint could mean less outside ventilation air that must be brought into the building, saving energy costs.

As a side note, do not put too narrow a definition on residential versus commercial equipment. In some cases, such a dividing line is mythical. There are some very good, very efficient "residential" products that fit perfectly well into a commercial environment.

Figure 3-2. A chiller at an HVAC plant. Courtesy: The Trane Co.

All-Air Systems

All-air systems use ducts to move warmed air from the central HVAC plant to the conditioned space. These offer the most control over zones, with less intrusion of equipment into the actual space except for the actual ductwork.

One initial drawback is the ductwork itself, which takes up space and capital costs for the installation, and requires some forethought when the building is being designed and constructed. It is difficult, sometimes impossible, to retrofit a building or space for ductwork if it has been built without it. In addition, air balancing becomes more difficult with larger systems, since air does not hold temperature well—some areas may be comfortable, while others farther from the mechanical equipment are too hot or too cold.

Air-and-Water Systems

Air-and-water systems are often used on building perimeters, or where exterior heat is needed more in the winter because of thermal transfer or leakage from windows and doors. Heat gain is also greater in the summer, mainly because of the windows, while interior cooling is called for throughout the year in almost all climates because of body heat and lighting loads. Air and water are cooled at a central HVAC plant before being distributed.

Individual zone control with this type of system is relatively easy through the use of thermostats. Smaller ducts and a smaller central HVAC plant can be used. Temperature setbacks can often keep the building warm with water alone, saving fan energy normally used to move air. Initial installation can be more expensive than an all-air system, and a second type of separate system is generally required for non-perimeter areas.

All-Water Systems

All-water systems transfer heat from water to the surrounding area, generally using radiators, baseboards or piping underneath the floor, or placed behind walls or ceilings-also known as hydronic heating. Pipes carrying water require much less space than air ductwork; more maintenance, however, is involved in these systems. Leaking pipes are more noticeable than leaking air ducts. Additionally, ventilation needs must be treated separately;

air has to be somehow moved into and out of a space whether or not the heating and cooling demands of the space require it.

Centralized Versus Decentralized Systems

HVAC systems are often divided into centralized or decentralized types. Most commercial office buildings have a centralized HVAC system. Rooftop chillers are mounted, of course, on the roof, while boilers are found in mechanical rooms, often in the basement so as not to detract from usable (or leasable) space. Centralized systems have a higher first cost but are more dependable and can be more economical in the long run. Heating and cooling is provided by a centralized HVAC plant.

Unit ventilators or fan coil units are used in many decentralized systems. These are found often in schools and hotels/motels. They are recognizable by their intrusion both into the occupied space and on the outside wall of the building. Each unit contains a coil for heating and cooling, and a blower motor to move the heated or cooled air. A great degree of control is its advantage. An

Figure 3-3. A boiler. Courtesy: Burnham.

unoccupied room can have this device turned off. A large number of these units, however, can add up to a rather large initial investment, and maintenance can pose a problem.

Many buildings use a combination of both. Most modern office buildings of sufficient size will have enough interior heat produced by people, lighting and office equipment to require mechanical cooling in interior zones in all but the coldest of weather. Perimeter heating is used in spaces nearest the building envelope.

Methods of Distributing Air and Heat

Methods of distributing air generally fall into constant air volume (CAV or CV) and variable air volume (VAV) systems.

Just as one might assume from their names, CAVs deliver a constant volume of air while varying the temperature of that air according to need. Conversely, VAVs rely on a constant temperature but vary the amount of air that is being moved.

VAVs, developed to cope with the energy crunch of the 1970s, were considered a great improvement over the older CAV systems. There was backlash when VAVs were blamed, somewhat unjustly, for many modern indoor air quality problems. But that criticism has been toned down and VAVs now are accepted as something we can live with, as long as they are properly maintained and operated.

Figure 3-4. A ceiling-recessed split ductless AC system. Courtesy: Mitsubishi Electric, Air Conditioning Products.

Unitary Equipment

Unitary equipment, a term one hears often in commercial HVAC, is divided into three categories according to use: residential, light commercial and commercial. General design in all three cases is the same, but specific designs and performance vary.

Unitary air conditioning units have a wide range of applications and have seen increasing use since they first began appearing in the 1930s. These differ from smaller, room-type air conditioners generally found in 5,000-14,000 British Thermal Units/hour (Btu/h) capacities (a Btu is a measure of heat).

The units consist of factory-matched components that are field assembled into complete systems to match end-user requirements. The equipment combinations include an evaporator (or cooling coil) and a compressor/condenser. These are often located on rooftops and cool the space by chilling water that is run through pipes in the building down into an air handler. Rooftop units frequently are the equipment of choice for cooling large commercial buildings. About three-quarters of a broad population—buildings from 30,000 sq.ft. requiring 75 tons of cooling, to 250,000 sq.ft. requiring up to 625 tons of cooling—is served by rooftop units. One advantage is they are preassembled and pretested, ensuring leak-free reliability when installed on-site.

Packaged Terminal Air Conditioners

Packaged terminal air conditioners (P-TACs) are another common air conditioning system for commercial applications, developed in the 1940s for through-the-wall use.

According to the Air-Conditioning Refrigeration Institute (ARI), a P-TAC consists of a wall sleeve with a separate unencased combination of heating and cooling assemblies specified by the builder and mounted through the wall. In this system's air-source heat pump, a heat exchanger passes inside heat to outdoor air when cooling and extracts heat from outdoor air when heating. P-TAC units are installed on an exterior wall and are primarily for commercial use to heat and cool a single room or zone.

P-TACs can be used for a small commercial building, or in abundance for a larger building such as a school, hotel or apartment building where there is no central air conditioning system.

Some types of businesses and buildings require more air

conditioning and more outside air than others. A fast food restaurant of only 2,500 sq.ft., for instance, can typically require 25 tons of rooftop cooling, and operate on up to 50 percent outside air. This is because a lot of that incoming air goes out again quickly to exhaust food cooking heat and odors.

Split systems, whereby the evaporator portion of the unit is inside while the condensing unit is on the outside, also are becoming increasingly popular because they are easy to install and do not require separate ductwork or piping.

Economizers and Heat Recovery

Most commercial cooling systems now come with economizers, which are vents or dampers installed in the return air systems. They will heat or cool the returned air, or bring in more outside air as needed—whichever is more economical. Frequently, particularly in northern climates early or late in the year, outdoor temperatures make it more economical to use direct outside air rather than reconditioned building air. This is fine in practice, but too much of a good thing has given economizers a bad name—they sometimes shut down either prematurely or completely to keep cold winter air out, reheating only building air—warm building air, but air that is also stale. An economizer should never operate

Figure 3-5. A packaged terminal system. Courtesy: McQuay.

fully closed to save energy, either in summer or winter. This would be starving a building of outside air. Better yet is to use a "smart" economizer, which will measure both temperature and humidity content of the air in determining the exact amount of outside air to use. Economizers cost from \$20-\$200 per ton of peak cooling capacity to operate, and are more effective in certain climates: energy savings can range from as low as 12 percent in a small building in the desert to up to about 80 percent in a temperature area such as San Francisco.

HVAC AND IAQ

HVAC and IAQ are inexorably related. Poor IAQ is often caused by poor ventilation, improper humidity, improper temperatures, unmaintained equipment or improper venting of areas that produce contaminants and odors—all concerns related to HVAC performance and design. The HVAC system may be a source of contaminants (examples: mold and mildew in a humidifier or duct, a dirty filter), or a pathway for contaminants (example: trucks outside the building are left idling near an air intake, which allows the air to enter the building). The road to better HVAC begins with a survey and analysis of the system itself as well as how it is operated and maintained. A more thorough on-site survey and analysis may be necessary if documentation on the original design is not available. IAQ-related health complaints, comfort-related complaints and occupant activities—desktop fans, tape or cardboard blocking supply air diffusers, space heaters, humidifiers and interference with thermostats (often accompanied by battles between employees over the perfect setting)—are all signs that the HVAC system is not meeting its responsibilities. In this event, the HVAC system may need to be modified (upgraded) or replaced.

Up to three general problems may be occurring when the HVAC system is not addressing the needs of the space:

1. The HVAC system's capacity and design cannot handle current needs. For example, suppose a space is renovated to become a kitchen area for employee use, but no exhaust fans

were installed to vent cooking odors. Or suppose a company elects to "green" itself by encouraging the use of indoor plants to make the working environment more pleasant, but excessive humidity in this case builds up to where people complain of "hot, stuffy" air and mold and mildew flourish. Or consider the building that includes extensive electronic equipment, building up more heat than the HVAC system's design capacity can remove to maintain comfort conditions.

2. The HVAC system is improperly operated and maintained. Control malfunctions, equipment problems, energy constraints, lack of manpower and lack of knowledge are all problems with operating and maintaining an HVAC system effectively. Suppose the maintenance staff is given an energy mandate to contain costs. The staff responds by resetting the outdoor air damper control below minimum design specifications. The building ends up underventilated whenever the outside temperature is extremely hot or cold. Or suppose the air intakes on the building envelope are not cleaned, and end up partially blocked by debris and bird droppings. The debris can be a source of bacteria, and can block air flow so that air balance is disrupted and negative interior pressurization results.

3. The HVAC system was never commissioned, so it may not be operating according to its original design specifications.

Specifically, we must examine the four major components of the HVAC function if we are to make a positive impact on IAQ, including the components below:

*   Ventilation (see Chapter 4). The ventilation rate should be based on ASHRAE 62-2001 guidelines of 15 cfm per person for general office space. The HVAC system should also employ an economizer when the economizer works to design, does not favor energy savings over IAQ, and is maintained and understood by building maintenance technicians. Air ducts should be checked for leakage. (See also Chapter 9 on mold, 2003.)

Table 3-1. Emission sources and problems identified in HVAC systems.

Source	Problem/Description
Seals, caulks, adhesives	Off-gassing of volatile organic compounds
Lubricating oils	Fans, motors in airstream
Ozone	By product of electrostatic air cleaners
Dust, skin	Dust mites
Organic debris	Insects, leaves, birds are or leave debris where microbes live
Cooling towers, drains, sumps	Microorganisms
VOC sinks	Dust and airborne contaminants
Cleaning compounds, disinfectants	Irritation
Boiler steam	Carries anticorrosives, chemicals that may wind up in airstream

- Filtration (see Chapter 5). The system should adequately filter contaminants out of outdoor air.

- Air balance (see Chapter 4). The HVAC system should provide a balance between air entering interior spaces and being exhausted back to the outside.

- Temperature (see Chapter 6). The indoor temperature should be maintained within a recommended range for comfort, depending on season and occupant activity.

- Humidification/Dehumidification (see Chapter 6). The system should provide supplemental humidification and dehumidification to maintain a range of 30-60 percent that is considered acceptable, with 50 percent being ideal.

Leaving out any one of these key ingredients creates an imbalance which is hard to remedy merely by increasing or enhancing the other three. When one is out of sync, it is hard to orchestrate the remainder of the chorus. For instance, moisture loads from outside air can often exceed the dehumidification capacity of the building's HVAC equipment, just as during the summer, cooling (and especially moisture removal) requirements can exceed HVAC system capabilities. Source removal, often considered the best way to rid a building of internal pollutants, sometimes is not possible. Bringing in more outside air to dilute these pollutants is another good idea, but can be expensive because we have to condition this air as well as dehumidify it. Filtration also helps, but not if the filter system is going to be easily overwhelmed by the load of pollutants we are trying to exclude.

Instead, using a combination of these efforts is necessary. Provide proper interior pressurization, exclude excess outside moisture through envelope modifications, and use dehumidification equipment where needed. These efforts will not only pay off in better air quality, but also in longer life for building materials and furnishings.

HVAC ENERGY SAVINGS BY DESIGN

Our lives are more dependent today on air conditioning than ever before. Besides the considerable heat imposed by the sun's direct rays and the effects of the sun warming the earth, consider the impact of people themselves, lighting and other electrical equipment. Electronic devices such as computers, fax machines and photocopiers have proliferated in recent years—all of these add to the heat needed to be removed, and add strain to the air conditioning system. Mechanical engineers 20 years ago could not predict the electronic revolution, and therefore many older HVAC systems cannot handle increases of heat without adding tonnage to the HVAC system, or without redesigning it completely.

In this problem lies an opportunity to recreate the HVAC system in an energy-efficient image, thereby reducing operating costs and reaping other benefits of modern technology. According to some studies, HVAC operating cost savings of $0.20-$1.00/sq. ft./year are possible by improving efficiency. If an IAQ prescription calls for increasing ventilation, which often requires adding capacity to the HVAC system, cost savings from use of energy-efficient components can, over time, offset the initial cost. Conversely, an HVAC upgrade can incorporate added ventilation with the added cost figured into the initial cost of the new system components and the return on investment calculations.

New products include adjustable speed drive controls, high-efficiency electric motors, high-efficiency electric fans and other new technologies. In fact, the National Energy Policy Act of 1992 mandates new minimum efficiencies for electric motors, favoring more efficient designs. The school of design itself is also approaching energy as a larger priority.

Table 3-2. Components of HVAC functions related to IAQ and applicable ASHRAE standards and guidelines.

HVAC Function	ASHRAE Standard or Guidelines
Ventilation	62-2001 "Ventilation for Acceptable Air Quality"
Temperature	55-1981 "Thermal Environmental Conditions for Human Occupancy"
Humidity	55-1981 "Thermal Environmental Conditions for Human Occupancy"
Filtration	52-1976 "Method of Testing Air-Cleaning Devices Used in General Ventilation for Removing Particulate Matter"
Commissioning	"Guideline for the Commissioning of HVAC Systems"

"If it ain't broke, don't fix it" is a motto often encountered in building maintenance programs. To time-beleaguered building operation and maintenance personnel, this translates to the very real: "If nobody's complaining, it's working fine, and I can work on many other problems that are priorities right now." But to facility managers and building owners alike, the economic benefits of energy-efficient designs are becoming an extremely important, and enticing, issue. And these gains are often related to productivity as well as the electric bill. For example, if upgrading the HVAC system makes it more efficient and allows it to provide a more comfortable environment, economic gains in both productivity and reduced energy consumption may be possible.

According to the Rocky Mountain Institute of Snowmass, CO, increasing productivity by even one percent is nearly equivalent to the entire annual energy cost of a building. This comes many years after the Building Emergency Thermostat Settings of the Carter era, which were a somewhat frantic approach to saving energy with resultant drastic reductions in the amount of outside air. At that time, few energy-saving alternatives were available other than "use less."

COMMISSIONING

Too many IAQ problems are the result of the simple fact the HVAC system is not meeting its design intent due to faults in design or installation. Wouldn't it be nice to have a form of insurance? That is what commissioning is all about.

One design engineer defines commissioning as, "The completion of one's job as a design engineer." Over the years, building systems have become more complex and varied, while the demands placed on them by owners and occupants have increased. Commissioning is a walk-through and test of the building's systems (not limited to HVAC) to ensure that the systems operate according to specification. In that sense, it is a quality assurance program for the building itself. To avoid commissioning is to have complete trust that all of the designers, engineers and contractors on the project did their job perfectly. Do not count on this happening.

ASHRAE Standard, "Guideline for The Commissioning of HVAC Systems" and guidelines from other associations can provide helpful generic commissioning roadmaps that can be tailored to any facility's requirements.

Sources

ASHRAE Handbook, HVAC Systems and Equipment, Atlanta: ASHRAE.

ASHRAE Handbook, Equipment, Atlanta: ASHRAE.

Ferreira, Al, "HVAC Benchmark Engineering," paper, IFMA Conference.

Haines, Roger W., "Ventilation Air, The Economy Cycle, and VAV," *Heating/Piping/Air Conditioning*, October 1994.

Hays, Steve M., Gobbell, Ronald V. and Ganick, Nicholas R., *Indoor Air Quality: Solutions and Strategies*, New York: McGraw-Hill, Inc., 1995.

Ingels, Margaret, *Willis Carrier, Father of Air Conditioning*, The Carrier Corp., 1991.

Langley, Billy C., *Major Appliances*, Regents/Prentice Hall, 1993.

Piper, James E., *Handbook of Facility Management*, New York: Prentice Hall, 1994.

Snips, "ASHRAE approves addenda to IAQ Standard," June, 2003.

Wendes, Herbert C., *HVAC Retrofits: Energy Savings Made Easy*, Atlanta: The Fairmont Press, 1994.

Chapter 4

Improving Ventilation

There are few standards or yardsticks we can point to in discussing indoor air quality. Air is seldom really "good" or "bad," nor is it easy to rate our indoor air on a scale of 1-10. When people get sick, or when people complain, that is one thing. Lack of unhealthy contaminants in our air is a goal worth shooting for, but what else? Proper ventilation is one of the few yardsticks we can use to describe healthy indoor air.

THE IMPORTANCE OF VENTILATION

Indoor ventilation is not some new idea or based on any new science.

Try this for historical perspective: Ventilation comes next to godliness, Edward P. Bates stated during his presidential address at ASHRAE's 1895 annual meeting. He said, "Every family has the right to have an abundance of good, fresh air, even if it is not aware of its rights. I hereby suggest that this be one of the first problems which we handle." In 1895, the Society adopted a minimum ventilation rate of 30 cfm per occupant.

In the intervening years, the number dropped to 10 cfm per person, then in 1973 ASHRAE revised Standard 62, which covers ventilation, to 5 cfm per person. No doubt, many building owners were relieved at the time, as the first energy crunch of the early 1970s signaled the end of a long-standing era of abundant and relatively cheap energy.

In 1987, the National Institute for Occupational Safety and Health (NIOSH) announced a study of 446 buildings with IAQ complaints that revealed "inadequate ventilation" to be a major contributor to the IAQ problem in 53 percent of the participating buildings. In 1989, ASHRAE published Standard 62-1989, which increased the minimum ventilation rate to 15-20 cfm per person, a substantial increase over Standard 62-1973.

Suddenly, many of the buildings constructed or renovated between 1973 and 1989, which included the many buildings erected during the office-building boom of the '70s, were and are considered underventilated. Increasing ventilation will certainly help toward solving most IAQ problems in underventilated buildings. But there are trade-offs, and all aspects of ventilation, just like all aspects of the HVAC function, must be addressed. In addition, the single most effective IAQ measure that can be taken is to remove or control contaminants. So is ventilation helpful? Yes. Is it a panacea for curing all sick buildings? Unfortunately, no.

VENTILATION AND IAQ

ASHRAE Standard 62 establishes guidelines for ventilation in various indoor spaces. Standard 62-1989 sets 15-20 cfm per occupant as the general minimum, unless the air is deemed healthful enough to justify a lower ventilation rate. Of particular interest to facility managers is that Standard 62-1989, at the time of writing, was in the process of being replaced with a newer, more comprehensive standard. The new, updated version, Standard 62R-2001, was scheduled with addenda in 2003.

ASHRAE 62-2001, with its higher minimum ventilation rates, came along at a time when fingers of blame were being pointed at faulty or inadequate building ventilation systems, as in the 1987 NIOSH study. Upon close inspection, however, the study points to a variety of contributors to poor IAQ related to building HVAC. Close inspection also tells us that the ventilation rate is only one factor to consider when addressing the total picture of building ventilation.

The NIOSH study was based on 446 buildings that had recorded complaints of upper respiratory illnesses and poor IAQ. Findings concluded that more than half of all problems were due to inadequate ventilation—not surprising, since 21 percent of the buildings investigated were operating with no fresh air at all.

Fifty-seven percent of the buildings had inefficient filters—15 percent of these because of improper installation—and 44 percent of the buildings had dirty ventilation systems. Surely, quantity of ventilation is important. But proper filtration and maintenance of the ventilation system are just as important.

Similarly, Healthy Buildings International (HBI), Fairfax, VA, reported that out of 813 building studies it carried out between 1980 and 1992, involving 750,000 building occupants, three-fourths (75.5 percent) were due to operating faults and/or poor maintenance. Fifty-four percent were due to poor ventilation prompted by energy conservation, 20.8 percent were due to poor air distribution, 56.9 percent due to inefficient filtration, and 12.2 percent due to contamination inside the ductwork.

It is clear from these studies that a flexible, holistic approach can be the best strategy for dealing with IAQ problems, incorporating ventilation as well as humidity control, temperature control, filtration and improved operation and maintenance.

In addition to addressing the HVAC system, of equal importance is controlling indoor contaminants. While we can rev up the system to address an IAQ problem, not controlling indoor and outdoor contaminants can be a costly mistake.

Note that the requirements of proper building ventilation are extensive and encompass the responsibilities of the designer, in-

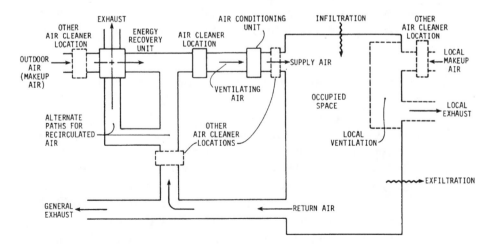

Figure 4-1. A typical ventilation system. Courtesy: ASHRAE.

staller and facility operator. In other words, quality indoor air cannot be left entirely up to the initial building plans or design, but is something that must be reconsidered and maintained over a period of time.

VENTILATION SYSTEM OPERATION

The ventilation system comprises a major portion of the HVAC system. A typical ventilation system is essentially made up of fans and ducts to supply clean air to interior spaces, and other fans and ducts to exhaust stale or contaminated air to the outdoors. In between, some recycled return air is mixed with outside air to supply the building, while the rest is exhausted. Also in between, the supply air is filtered, humidified, dehumidified, heated or cooled by components in the air handler, depending on the capabilities of the system and the demands of the season.

Mission of the Ventilation System
The mission of an effective ventilation system's design is to:

1.   Dilute indoor contaminants and concentrations of carbon dioxide (CO_2) emitted by breathing.

2.   Distribute air evenly to ensure all occupants have access to fresh air that has relatively uniform temperature, humidity, air velocity and quality.

3.   Maintain a balance of air pressure between air indoors and outdoors.

Operating Components and Characteristics
In a typical ventilation system, outside air flows through an intake into a damper. These can be manually or automatically operated. They can be closed to prevent the spread of smoke in a fire, but closing inadvertently can restrict the flow of outside air. The dampers should be adjusted as part of the overall system balance and should not be changed arbitrarily. The amount of air allowed in is determined by controls.

The outside air enters an air-mixing plenum, where it is combined with recycled indoor air. The resulting mixture is called supply air. It enters the air handler, where it is filtered, heated, cooled, humidified and/or dehumidified. The supply air is then delivered to the interior spaces via ductwork, usually located in the ceiling, then distributed with supply air diffusers. Air grilles and diffusers are where dirt usually becomes trapped before entering the space, and can be a good indication that something is wrong with the filtration system. Also, their placement is important. They need to deliver air where it is needed, near the occupants. The layout of the space may have changed since the original installation plans for the dampers and grilles were laid.

Ideally, at a given ventilation rate air flows through interior spaces, where it provides thermal comfort, dilutes contaminants and supplies fresh air for occupant use.

The air is then pulled by fans through return air grilles, where it becomes return air. Return air flows through the return air plenum, where some enters the air-mixing plenum to repeat the loop, and some is exhausted to the outside via vents, including

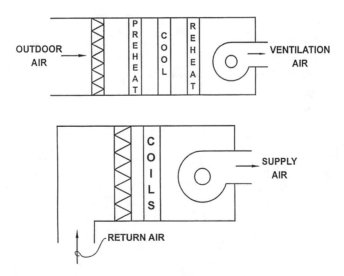

Figure 4-2. Basic components and function of a ventilation system. Outdoor air is conditioned and distributed to interior spaces. Air is then exhausted, some to the outside, some back into the system for reconditioning and distribution.

through bathroom exhaust vents, kitchen fans, etc. Of course, some natural mixing also occurs when interior air is lost through opened doors, around any leaking windows or foundation cracks, etc.

Elements of Ventilation

The success or failure of the ventilation system rests with a number of interdependent components and principles, including airflow, use of return air in supply air, demand control, air balancing, air distribution and filtration.

Airflow - This is the amount of clean indoor air introduced at a given ventilation rate. The ventilation rate is measured in cubic feet per minute (cfm). There are several ways of viewing cfm:

The term "cfm of outdoor air" refers to the total amount of outside air entering the building, in cubic feet per minute.

The term "cfm per person" refers to the amount of ventilation

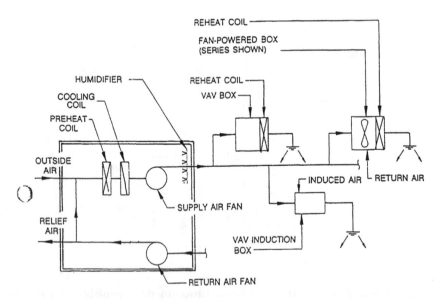

Figure 4-3. A variable air volume HVAC system consisting of a centralized air handler with various VAV terminal units. Courtesy: ASHRAE HVAC Handbook.

per occupant in the space, thereby dividing the total quantity of outside air by the number of occupants. It is based on the maximum number of occupants who could be present in the space at any time.

The ventilation rate can also be calculated in cfm per sq.ft., thereby dividing the total quantity of outside air by the square footage of the affected space.

Finally, the ventilation rate may be expressed in number of air changes per hour.

A suitable ventilation rate should be selected based on current ASHRAE Standards or other national or local building code guidelines, usually 15-20 cfm per occupant at present.

Percentage of Outside Air to Supply Air - In typical systems, some return air is sent back through the ventilation loop, while some is exhausted back outside. Outside air and return air are combined in an air-mixing plenum. The percentage of this mixture that is outside air is called the "percentage of outside air to supply air."

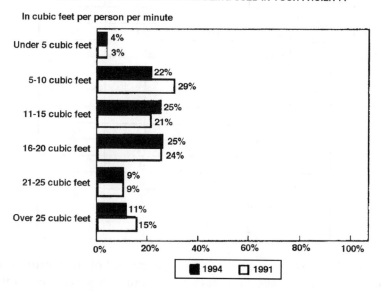

Figure 4-4. Representative sampling of ventilation rates, based on a study conducted by IFMA. Courtesy: IFMA.

Recycling indoor air can be an important strategy if the building owner wishes to increase airflow without inflating energy costs. While ASHRAE Standard 62 favors introduction of outside air, it allows for increased use of recycled indoor air if the outside air is too polluted, or to conserve energy. For this strategy to be successful, the return air must be properly filtered to avoid

Air Changes per Hour	Minutes Required to Achieve Removal Efficiency of:		
	90.0%	99.0%	99.9%
1.0	138	276	414
2.0	69	138	207
3.0	46	92	138
4.0	35	69	104
5.0	28	55	83
6.0	23	46	69
7.0	20	39	59
8.0	17	35	52
9.0	15	31	46
10.0	14	28	41
11.0	13	25	38
12.0	12	23	35
13.0	11	21	32
14.0	10	20	30
15.0	9	18	28
16.0	9	17	26
17.0	8	16	24
18.0	8	15	23
19.0	7	15	22
20.0	7	14	21
25.0	6	11	17
30.0	5	9	14
35.0	4	8	12
40.0	3	7	10
45.0	3	6	9
50.0	3	6	8

Figure 4-5. Air changes per hour and the amount of time, expressed in minutes, it would take to remove 90, 99 and 99.9 percent of airborne contaminants. Courtesy: Healthy Buildings and United Test & Balance Service, Inc.

reintroducing contaminants into the building spaces (see Chapter 5, and mold, Chapter 9).

Demand Control - This term is associated with the process of allowing demand for fresh air to determine the number or frequency of air changes. The trigger for another air change is often a monitor or sensor that measures the amount of CO_2 in the space emitted by human breathing. Once CO_2 builds up to a certain level (this level is still under debate, but is in the range of 800-1,200 ppm) an air change will take place.

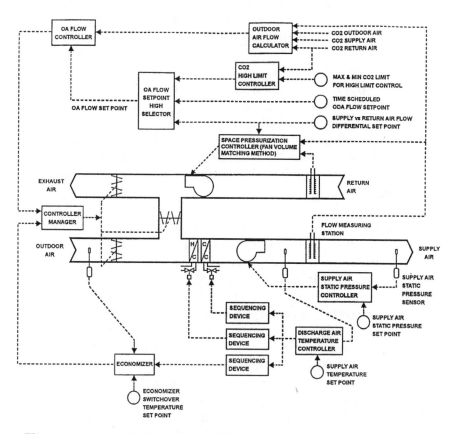

Figure 4-6. A sophisticated variable air volume HVAC system. Courtesy: ASHRAE, from the *ASHRAE Journal*.

Air Balancing and Pressurization - These terms refer to distribution of air in the conditioned space and air pressure within the building envelope. A negative pressure can result in conditions where unconditioned air enters the building through all available seams and cracks to correct the imbalance on its own. The goal is to maintain a proper air balance.

Positive Pressurization. If more air enters the building than exits via the exhaust vents, then positive pressurization will result. In other words, the pressure inside the building will be slightly higher than it is outside. Positive pressurization will cause the excess amount to exit the building through all available seams and gaps in the building envelope. Slightly positive pressurization is considered a desirable condition because it helps to keep out cold or hot outdoor air as well as moisture, dust and contaminants. Excessive positive pressurization will cause heating and cooling dollars to be wasted.

Negative Pressurization. If more air is leaving the building than entering through the outdoor air intake, then negative pressurization will result. This will draw air into the building to correct the imbalance. Unfortunately, the air entering is not heated, cooled, humidified or dehumidified, which can affect humidity levels and thermal comfort (and energy costs). In addition, the air may be bringing in contaminants from loading docks, nearby traffic or other sources.

Air Distribution - Clean indoor air must be distributed evenly in indoor spaces so that all occupants have access to fresh air of relatively uniform thermal comfort, humidity, air velocity and quality. In addition, perimeter areas should be properly ventilated during the winter (heating season) to keep these areas warm and dry to prevent condensation. The resulting moisture can damage walls, wall-coverings and drapes, while providing conditions that allow mold and mildew (harmful bacteria) to flourish (see Chapter 6 for more on moisture and humidity).

Regarding thermal comfort, two typical problems with air distribution include drafts and temperature gradients. Drafts result when air is unevenly distributed and occupants complain of being cold in certain areas or on certain body parts. Temperature gradients are prevalent when occupants complain that a room feels warm near a radiant heat source (such as a baseboard heater) but feels progressively cooler the farther they are from the heat source. Temperature gradients can also be vertical, as when warm air enters the space from a supply air diffuser in the ceiling, but hangs near the ceiling and does not reach the floor (see Chapter 6).

Filtration - Filtering supply air is vital to removing particulate contaminants and sometimes odors. Typical filter problems include clogged or wet filters. Solutions include scheduled maintenance of filters, the specification of filters that are more efficient and the introduction of air cleaning devices (see Chapter 5).

TYPICAL VENTILATION PROBLEMS

Lack of ventilation can lead to IAQ problems when:

1.   The system is acting as a source of contamination.

2.   The system is acting as a pathway of contamination.

3.   When ventilation deficiencies allow for a buildup of contaminants.

The ventilation system may be acting as a *source* of contaminants when:

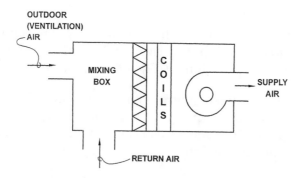

Figure 4-7. Simple depiction of a constant air volume HVAC system.

- Microbiological contamination is prevalent in higher-moisture areas such as ductwork, humidifiers, dehumidifier drain traps, outdoor air intakes and other components.

- Contaminants are collecting on supply air diffusers and return air grilles.

- Debris, such as bird droppings, may be lodged in the outdoor air intake, acting as a source of microbiological contamination.

The ventilation system may be acting as a *pathway* for contaminants when:

- All of the above contaminants located in the ventilation system are using the ventilation system as a pathway.

- Outdoor air intakes may be located near sources of moisture or contaminants, causing the ventilation system to act as a pathway for the contaminants—e.g., pesticides or lawn cuttings entering an outdoor air intake that is located on ground level—that are overwhelming the filtration.

Ventilation deficiencies can occur when 1) space needs have changed but the ventilation system has not been properly modified to address the new needs, 2) the original design of the system is insufficient in bringing in enough outside air, or 3) the ventilation system is improperly operated and maintained.
Ventilation deficiencies include:

- Inadequate ventilation—the ventilation system is not delivering sufficient volume of clean indoor air to dilute indoor contaminants and dilute CO_2 concentrations.

- Outdoor air intake controls have been adjusted to tighten airflow due to an energy conservation program, restricting the amount of outside air entering the building and resulting in inadequate ventilation.

- Debris is blocking the outdoor air intake, restricting airflow.

- Ducts are damaged, restricting or leaking airflow.

- Negative pressurization resulting from blocked airflow, possible causes including debris blocking the outdoor air intake, poor design, leaking ductwork on the supply end or problems resulting from an unducted return air plenum.

- Excessive positive pressurization resulting from poor design or other causes.

- Poor air distribution.

- Improper filtration—filters are inefficient compared to the building's needs to remove particulate contamination, are microbiologically contaminated due to moisture, or are poorly maintained.

- Components such as dampers or controls are malfunctioning.

CORRECTIVE STRATEGIES

There are a number of options available to the facility manager to improve the ventilation system or ensure a new ventilation system meets IAQ needs. Specific choices will depend on unique building conditions and the advice of qualified professionals in IAQ, HVAC and/or mechanical engineering.

First, picture the ventilation system as a tube shaped in a U with one end to introduce air into a box and the other end to exit air out of the box. Our basic strategy is to:

1. Ensure that a sufficient volume of air is passing through the tube.

2. Ensure that no space inside the tube is blocked to disrupt air balancing and restrict airflow.

3. Ensure that the space inside the tube is clean and dry to prevent the growth of mold, mildew and bacteria (See Chapter 9).

4. Ensure that air is not leaking from the tube in any location.

5. Ensure that the opening at the air-introduction end of the tube is not blocked or located too close to any sources of contamination.

6. Ensure that air is distributed evenly in the box and reaches all occupants.

This is an oversimplification, of course, of the steps needed to ensure the ventilation system is supporting our IAQ objectives, but it is helpful in that it defines a scope of activity. Specific measures include:

1. Check current ventilation rate against current ASHRAE guidelines and increase ventilation rate as needed to match guidelines. Conduct an analysis to ensure that the ventilation system is sufficiently meeting current occupant needs based on current use of the building. Check past occupant complaints regarding humidity, thermal comfort and "draftiness." Test the ventilation system's ability to reach the occupant level and distribute air uniformly at the occupied level. Check air balancing to ensure proper pressurization.

2. Specify filters with a higher efficiency and clean and replace them regularly, while considering the use of additional air cleaning devices. The less outdoor air that is introduced, the cleaner must be the indoor air. Filtration becomes more important than ever in these cases, along with control of pollutants or contaminants that filters are not able to cleanse. Storing gallons of poorly-sealed cleaning fluid, for example, will not be helped much by the best building filtration system. Filters are of no use, for example, against radon, but can be very effective in filtering out particulates—especially normal amounts of dust and dirt.

3. Use charcoal filters where appropriate to remove odors from either interior or incoming air (see Chapter 5).

4. Add wire mesh screens ("bird screens") to outdoor air intakes to prevent debris from entering the outdoor air plenum.

5. Ensure that the outdoor air intakes are located sufficiently above street level if facing a street, and a suitable distance from any building exhaust, loading docks, dumpsters, or combustion gas emission sources (for the latter, some building codes require a minimum distance of 25 ft.).

6. Relocate outdoor air intake away from sources of contamination, or remove or move source of outdoor contamination, denying the source a pathway into the building, if contaminants are overwhelming the filtration buffer.

7. Check the air handler, ducts, plenums, supply air diffusers, return air grilles and other components for moisture, debris, mold, mildew and undesirable levels of dirt and dust; keep ductwork and other ventilation components, as well as HVAC components in air handler, clean and dry (see Chapter 6).

Figure 4-8. Typical dampers and controllers used in ventilation systems. Courtesy: Honeywell, Inc.

Figure 4-9. Typical louvers used in ventilation systems.

8. Add ductwork to return air plenum to support pressurization balance and prevent dirt, dust and microbiological contaminants from being carried back into the system in the return air.

9. Check fans for undesirable dirt build-up, moisture and microbiological contamination, and keep clean and dry.

10. Check all controls to ensure their proper operation.

11. Check ductwork for damage and leaks and repair as needed.

12. Tighten the building envelope to prevent undesirable outdoor air from entering the building. Check to see if natural ventilation—e.g., windows—is allowing contaminated air, from local traffic, for example, to enter the building.

13. Specify special supply air diffusers to improve air mixing and thereby improve air distribution and ventilation effectiveness. A good diffuser will enhance air mixing, providing air where it is needed while eliminating some of the excessive cost of heating and cooling. Use of specialized diffusers can be a means of providing zone control without tinkering with the original mechanical equipment or ductwork in a building.

SUPPLY AIR: THE DEBATE OVER OUTSIDE VS. INDOOR AIR

As stated above, if we increase airflow to introduce more fresh air and do a better job diluting contaminants indoors, the air can come from outside or a mixture of outside and inside air. Air mixing takes place in the air-mixing plenum attached to the supply side of the air handler, where air is filtered and conditioned in typical air systems. The amount of indoor air that is recycled depends on the quality and capabilities of the system's air filters (as well as on the condition of the rest of the system, such as the return air plenum). Theoretically, clean, recirculated air in a building's interior could make up part of the influx of air needed to keep a building well-ventilated throughout the day.

ASHRAE 62 tends to favor the introduction of outside air. As ASHRAE ventilation standards undergo revision, so does the debate over the amount of outdoor versus the amount of indoor air. Critics contend that relying solely on outdoor air in order to increase ventilation will not only increase energy costs (as outside air takes more work and therefore more electrical energy to condition than indoor air), but may introduce heavier contaminants that are located in the air or are being produced on-site. There are also certain cases where recirculated air is actually better for building occupants than outdoor air—such as in highly polluted cities or in cases where demand for quantities of clean air is more than what the "natural" environment can provide—hospital operating rooms or manufacturing clean rooms, for example. Critics also contend that adding more outside air fails to take into account recent improvements to filtration. A decision here could mean the difference in thousands of dollars in energy and operating costs at a facility over several years.

One thing is certain: Increasing the amount of ventilation using only outdoor air costs more. Costs for installation of a supplemental system to heat and cool outside air averages $6.25/cfm for cooling and $2.00/cfm for heating or a total cost of $8.25/cfm. In the humid South, the latent heat loads (removal of moisture in the air) have the potential to double cooling costs in systems using only outdoor air circulation.

The Clean Air Device Manufacturers Association (CADM) believes that acceptable air quality can be achieved through the use of better filtration systems that would reduce the amount of outside air exchange to only 5-10 cfm per person. If this were true, the savings per year would be formidable.

The installed cost of air-cleaning systems will vary from under $5.00/cfm for small room area packaged units to under $2.50/cfm for large central station air-cleaning systems. Operating costs for the use of an air-cleaning system for one year is only about $0.40/cfm compared to conditioning outside air, which averages at $1.40/cfm.

There is another way to get around providing more outside air. This is by having a building that does not need it—in other words, one that does not have high internal pollutant levels. The standard refers to this possibility as an "alternative ventilation rate procedure." What constitutes high pollutant levels is still in many cases a mystery.

Use Common Sense

The demands of any ventilation standard can be seen as expensive and maybe even overly wasteful only if we do not temper the requirements with common sense and personal observation. Remember that standards are really just recommendations unless implemented as code or law.

If a space is to be renovated or constructed, therefore, the Standard must be followed if implemented as code or law; but for an ongoing IAQ program, the facility manager has a lot of latitude.

Consider a school auditorium with a capacity of 500 people

designed in 1979. Designed using the old Standard, 5 cfm per person, or 2,500 cfm of 53°F supply air, is provided. Under ASHRAE 62-2001, with 15 cfm per person required, the total ventilation would need to be increased to 7,500 cfm.

Could a lesser amount of outdoor air be realistically applied to this situation than the full 7,500 cfm, without risking inferior air quality and occupant complaints? One question might be, how often is this room used, what are the average lengths of time it is in use, and what are the possible sources of contaminants?

We will assume there is no smoking allowed. People then become the logical pollution source: CO_2, moisture and body odor. Significant energy costs could be avoided by finding a figure somewhere between 5 and 15 cfm per person rather than increasing to 15 cfm per person.

THE VALUE OF PROPER AIR DISTRIBUTION

The solution to pollution may be dilution—in some cases. But if we consider only the gross volume of air movement through a building, or even through each sector of a building, the ventilation standards may be achieved without actually delivering fresh air to many of the people. This results in the common office workplace stressor, "lack of air movement," and can allow indoor pollutants to accumulate to harmful concentrations. It can be a big problem and not a single dollar should be invested in an IAQ improvement related to ventilation unless it is considered.

According to author and IAQ expert Phil Bearg in his book, *Indoor Quality and HVAC Systems*, published by Lewis Publishers of Boca Raton, FL:

> The introduction of adequate quantities of outdoor air into the HVAC equipment is a necessary, although not sufficient condition for the maintenance of good IAQ. Not only must enough outdoor air be drawn into the system, but this outdoor air for ventilation must also be delivered to where the building occupants are [located] ... The techniques for determining the [outdoor air] quantity entering an HVAC sys-

**tem include either a direct measurement of the volume of
this air stream, or measurement of the total air volume in
combination with a determination of the portion of this total supply air volume that is outdoor air.**

If the HVAC system is not properly designed, the fresh or
treated air can "short circuit" directly from the supply diffusers
to the return grilles, without properly mixing with the ambient
air in the occupied zone. This phenomenon happens frequently
on the outside of the building as well, where an exhaust vent
will be located too close to the supply vent. Some building codes
spell out how far apart these must be, but often there is no exact
measurement required on this. Temperature inversions, weather,
barometric pressure, etc. can all play a role. In some cases, we
can eyeball the vents and tell just by looking if they are too close
together.

Here is a tip: A smoke tube can also be used in many simple
ventilation inspections. This is a common, inexpensive tool that
usually does not use actual fire or smoke (some do, however) but
more commonly an aerosolized powder that draws moisture
when released and gives the appearance of smoke to indicate air
movement. If smoke released from one of these near the exhaust
vent drifts over to be drawn into the building's intake, we know
immediately there is a problem. Of course, we may have to perform this test on more than one occasion, or do it on a day when
there is little wind. In some cases, the problem can be easily cured
by adding a sheet metal extension to one or both vents, or adding
some sort of barrier or screening material. In other cases, there is
little we can do without incurring massive expense and relocating
the vents. Consult a mechanical engineer on that matter.

Without a design that ensures proper mixing, people can be
breathing stale air, while fresh air is passing over their heads or
otherwise bypassing the air mass they are actually breathing. The
degree to which fresh air intermixes with ambient air in the occupied zone reflects the ventilation effectiveness of the system.
Again, diffusers can play an important role. Many ceiling diffusers
do little more than acting as cosmetic screens to cover up the
supply air vent. Some of the better designs will actually aid in
mixing incoming air into the available space.

**Air Distribution
Is An Ongoing Commitment**

Standard 62-2001 alerts building designers to potential source problems for makeup air, warning against picking up cooling tower drift, engine and kitchen exhaust. Another section of the standard emphasizes the need to monitor the amount of outside air being brought into the system. To do this requires a way to accurately measure actual airflow into the system as constructed, rather than simply as a design flow. Although continuing to measure the amount of airflow after HVAC system start-up is not a requirement, facility managers will need to be aware that system airflows can and often do change over time, either from system modifications or from variations in the static ventilation head because of changing building characteristics. Regular measurement and adjustments will be necessary to assure that the building continues to meet minimum airflow recommendations.

Office Partitions

Some IAQ experts argue that office partitions are impediments to proper air mixing. Makers of office partitions, however, believe this need not be the case, arguing that heat loads and positioning of air diffusers more significantly impact ventilation effectiveness than do airflow gaps on diffusers or partition heights. One company, Steelcase, produced a report which stated that air flow panels built into the lower portion of partitions show little benefit as long as temperature ranges are comfortable and air flow diffusers are positioned correctly.

Of course, when barriers are frequently moved without thought to room air circulation, a problem can arise. Office workers often come up with their own makeshift cardboard diffusers taped to the ceiling vents in order to block or redistribute air that is too cold or too warm, or airflow that is simply considered annoying. IAQ inspectors spotting these on a walk-through inspection know there is a problem with air circulation or with the supply air temperature.

Careful integration of environment configuration with HVAC design and deployment provides the most effective approach to ensuring high-quality circulation of air.

ASHRAE STANDARD 62-2001

ASHRAE Standard 62-2001, "Ventilation for Acceptable Indoor Air Quality," establishes a definition for acceptable indoor air quality as "air in which there are no known contaminants at harmful concentrations as determined by cognizant authorities and with which a substantial majority (80 percent or more) of the people exposed do not express dissatisfaction."

Standard 62-2001 then establishes a process for reaching this goal in typical indoor spaces using either the Ventilation Rate Procedure or the Indoor Air Quality Procedure. The Standard sets 15-20 cfm per occupant as the general minimum, unless the air is deemed healthful enough to justify a lower ventilation rate. That is where the Indoor Air Quality Procedure comes in.

Ventilation Rates—A History

Establishing minimum ventilation rates has long been a by-guess-and-by-gosh proposition. While ancient buildings had plenty of indoor/outdoor air changes, this was largely a by-product of looser construction building practices and materials. It was not until the 1700s that chemists and physicians began to associate ventilation with preventing the spread of disease and healthy air to breathe. In the early 1800s, 5 cfm of fresh air per person was considered acceptable. The first major change in this thrust took place in 1895, when the American Society of Heating and Ventilating Engineers (ASHVE), founded in 1894, came up with a standard of 30 cfm per person. This became law in many states over the next few years. This increased ventilation rate required the work of engineers, and mechanical ventilation systems came into vogue for the first time. It was not until the 1940s that this standard began to erode. A minimum 5 cfm per person was adopted in ASHRAE Standard 62-1973. However, this was suggested only as a minimum and was expected to be used with some amount of supplemental recirculated room air. It was not until 1975, when the Arab oil embargo was in effect and the energy crisis loomed nationally, that minimum ventilation rates of 5 cfm per person really began to be accepted—at the same time that buildings were being sealed up to decrease drafts or leakage. This ventilation standard stood until 1981, when it was amended and increased in certain cases. However, in many areas the earlier standard remained in effect. It was not until 1989 that the standard was increased to 15-20 cfm per person. The 2001 changes the standard for optimal indoor air quality for new and existing buildings, setting for minimum ventilation rates. Addenda reported in 2003.

The Ventilation Rate Procedure

The Ventilation Rate Procedure tells us that acceptable air quality is achieved by providing ventilation air of the specified quality and quantity to the space. Whenever the Ventilation Rate Procedure is used, the design documentation should clearly state that this method was used and that the design will need to be re-evaluated if, at a later time, space use changes occur of if unusual contaminants or unusually strong sources of specific contaminants are to be introduced into the space. If such conditions are known at the time of the original design, the use of the Indoor Air Quality Procedure may be indicated.

The Indoor Air Quality Procedure

In the Indoor Air Quality Procedure, acceptable air quality is achieved within the space by controlling known and specifiable contaminants to accepted levels. The Indoor Air Quality Procedure could result in a ventilation rate lower than would result from using the Ventilation Rate Procedure, but the presence of a particular source of contamination in the space may result in a higher rate. Change in space use, contaminants, or operation may require a re-evaluation of the design and implementation of needed changes.

Comparison of Procedures

The Ventilation Rate Procedure is generally characterized as the more practical or easier to use. While sometimes it is difficult to measure airflow, it can be done using modern instrumentation and careful measurement practices. The Indoor Air Quality Procedure, however, involves measuring varying levels of multiple contaminants—if there are no problems, the overall ventilation rate is assumed to be acceptable, and need not be increased. The problem is that measuring those various levels and types of contaminants can be even more difficult to do than measuring ventilation rates.

Determining Space and System Ventilation Requirements

1. Qualify outdoor air quality. A table in the Standard lists air quality criteria for outdoor air as set by the U.S. EPA. Based on these criteria, the designer must evaluate the building site and provide whatever filtration is required to assure that outdoor air brought into the building is acceptable for ventilation. In some instances, it may even be necessary to reduce outdoor airflow during critical periods.

2. Look up the prescribed ventilation airflow ("outdoor air requirement" or OAR) for each space in Table 2 of the Standard. Table 2 gives ventilation requirements for 81 different space types. Most ventilation requirements are expressed as "airflow (cfm) per occupant" since contaminant generation is presumed to be proportional to the number of people in the space. Rooms provided with local exhaust (e.g., rest rooms, kitchens, smoking lounges) may be ventilated with air supplied from adjacent spaces.

3. For each space, solve the equation $V_o = OAR \times OCC \div E_v$, where:

—V_o is the required volumetric flow rate of outdoor air
—OAR is the applicable Table 2 value (expressed either as cfm per person or as cfm per sq.ft.)
—OCC represents space occupancy (usually the maximum design occupancy)
—E_v is ventilation effectiveness—i.e., the fraction of ventilation air delivered to the space that actually reaches the occupied zone. For example, if E_v is 0.80, 80 percent of the air supplied to the space enters the occupied zone while the remaining 20 percent bypasses it.

It is important to understand the distinction between "occupied zone" and "occupied space." While the floor, walls and ceiling define the "occupied space," people are in the "occupied zone"

(Continued from page 106)

which is 3-72 inches above the floor and more than 24 inches from the walls. Ventilation airflow is required for the occupied zone, not simply for the occupied space. Note, too, that a space is deemed intermittent occupancy if it is continuously occupied at maximum capacity for less than three hours. In such cases, space occupancy for the purposes of this equation can be the average occupancy, but not less than one-half of the maximum design occupancy.

4. Determine the total outdoor airflow required at the air handler to satisfy system ventilation needs. In a system with a dedicated supply air source per space, the required system-level outdoor airflow equals the sum of the outdoor airflow required in each space.

    To determine the system outdoor airflow requirement for a multiple-space system, one with recirculated return air and a central air handler serving more than one room:

    a) Calculate the uncorrected ventilation fraction, X, by dividing the sum of the space ventilation airflows (V_{on}) by the sum of the space supply airflows (V_{st}): $X = V_{on} \div V_{st}$.

    b) Identify the "critical space": that's the space that requires the highest percentage of ventilation air at minimum supply airflow. The critical space has the highest space ventilation fraction; that is, required outdoor air flow (V_o) divided by minimum supply airflow or Z.

    c) Use equation 6-1 of the standard, $Y = X \div (1 + X - Z)$, to calculate the corrected ventilation fraction, Y—i.e., a value that is greater than X but less than Z.

    d) Compute the actual peak system outdoor airflow required by solving the equation $V_{ot} = Y \times V_{st}$.

Consider the example below of how to meet the requirements for Standard 62-2001. This example is for a single-space constant-volume system, a classroom designed to accommodate up to 50 students but with an average of 30, considered intermittent occupancy.

(Continued from page 107)

1. Determine airflow needs. To provide the required ventilation, first determine how much outdoor airflow the classroom needs. Begin with the Ventilation Rate Procedure. Table 2 of the standard indicates that the minimum outdoor airflow rate for classrooms is 15 cfm per person. Assuming that the ventilation effectiveness is 1.0—i.e., that 100 percent of the air supplied to the classroom actually enters the occupied zone—the classroom requires 15 cfm per person × 30 people, or 450 cfm.

    The Indoor Air Quality Procedure could be used if we are comfortable enough to make assumptions about furnishings and contaminant sources. If we assume all furnishings emit low levels of VOCs, and no other noxious contaminants exist, then the CO_2 level must not exceed 1,000 ppm. If we assume the outdoor air contains 350 ppm of CO_2 and that each occupant generates CO_2 at the rate of 0.011 cfm, then the classroom needs 507 cfm of outdoor air for compliance—i.e., N ÷ Cs – Co (0.011 cfm per person × 30 occupants divided by [0.001000 - 0.000350 parts by volume]), where:

—N represents the contaminant generation rate
—Cs and Co represent the volumetric concentration of the contaminant in the space and outdoors, respectively

2. Determine airflow. Regardless of the method used to determine space ventilation needs, the next step is to discover how much outdoor airflow the system requires. Since there is only one space in this example, system and space ventilation requirements are identical. Compliance with the standard demands 450 cfm via the Ventilation Rate Procedure or 507 cfm via the IAQ Procedure.

(Changes may be added with ASHRAE Standard 62-2001 with addenda in 2003.)

Source: Dennis Stanke and Brenda Bradley, *Engineers Newsletter,* **The Trane Company.**

A Word on Engineering and Technical Standards

Most of the standards we refer to here, including ASHRAE Standard 62-2001, are really only recommendations. Standards by themselves are not law or legally binding. But these standards are often adopted at least in part by local or regional building codes, in which case they must be followed as law. They can also be brought up in a court of law, if things get to that point, to be used by lawyers on behalf of their clients to determine whether the ventilation or other engineering, design and construction practices used in a particular building were adequate or not according to a consensus of professional standards at the time. Mark Diamond, a New York attorney who specializes in indoor air cases, points out that indoor pollution "is exacerbated by the failure of most buildings in America to dilute stale indoor air with comparatively cleaner outdoor air." He points to the growing liability that employers and real estate professionals face. "Increased public attention to indoor pollution, combined with growing recognition by the medical and legal professions, will lead to more lawsuits," he said.

The New Standard 62

As noted earlier in this chapter, Standard 62-2001 is in the process of revision and will be released in the 2002- time frame. For now, we will call it "Standard 62R."

Standard 62-2001, like those before it, was the work of a committee, subject to extensive peer review, input and comment before it was adopted. The newer standard goes considerably beyond ventilation. It was because of this that the technical committee reviewing this standard appeared to balk at its implementation. A careful review of the entire proposed Standard by the committee at an early stage showed that the methods proposed in several sections of the standard were overly complex, unworkable, burdensome to designers and most important of all, unnecessary—the proposed Standard forced designers to be responsible for anticipating all possible uses of spaces, materials selection, all possible sources of emissions and obtaining emissions data from the manufacturers. This was beyond the traditional responsibility and expertise of the designers. It was pointed out that this unrealistic situation could have serious legal implications and was of great concern to members of the technical committee.

Many of these concerns were later remedied, just as any standard goes through plenty of professional debate before it is accepted. But the level of concern here was well-founded.

Considering that five tons of HVAC equipment capacity is required to cool every 1,000 cfm of outside air, the committee working on this standard was faced with the enormous and difficult task of setting adequate ventilation rates without brutally penalizing building owners with high energy costs.

Other significant aspects of the Standard 62R:

- Access to ventilation systems will be called for, making cleaning easier later on.

- Internally lined ductwork will be discouraged if relative humidity in the ductwork is above 60 percent, to discourage microbial growth from moisture build-up.

- Reference to CO_2 demand-based ventilation systems.

- More sophisticated controls will be necessary to bring in more outside air even after temperature has been satisfied.

- Ventilation systems will have to be turned on prior to building occupancy for the day, to head off pollutant build-up.

- Recognition of the importance of relationship between the building mechanical system and the thermal envelope—not typically a part of the architect-engineer relationship.

Why a New Standard is Needed

Often, it is only after a standard is adopted does it begin receiving widespread attention, at which point it is open to criticism and perhaps, later on, revision. These are working standards, just as our buildings and facility management systems and practices are working and evolving, not static. While those studying such a standard do their best, sometimes real-world situations prove that the standard may have had unseen flaws, or shortcomings. Additionally, committee members may have shown their own professional or personal biases. In some cases, standards are adopted over the objections of certain groups who will continue to lobby their cause(s) until a new or revised standard comes along. This is a laborious and time-consuming process. Meetings to review the standard or proposed revisions must be held when members of the committee can attend, at a location and time they can hopefully all be satisfied with. While there is no certain prescribed amount of time allotted to coming up with a new standard, the ASHRAE process generally takes about eight years.

Differences, Disparities in Standards

A number of difficult questions about the new Standard 62R have been posed. For one thing, ventilation is seldom the one-stop solution to problems with tight buildings. At best, it may be an expensive solution. Open a window in an air conditioned house in the summer and what happens? We lose much of the air we have spent money to cool.

Similarly, John Bower in his book *Understanding Ventilation* makes a strong case for improved ventilation for houses—changes

Figure 4-10. The air handler shown is specially designed for IAQ sensitivity, featuring insulation with a cleanable acrylic coating incorporating an EPA-registered antimicrobial that resists growth of bacteria and fungi. Courtesy: Carrier Corporation.

which have a commercial appeal as well. Natural and "accidental" pressures seldom supply occupants with enough fresh air, he states. It is often either too much or too little, or not distributed to where it is needed the most. Bower, a "healthy house" advocate, points to the limits of filtration effectiveness and the higher costs that will be incurred. However, ventilation is a rather poor response to smoking—because such vast amounts are needed to dilute the smoke. We also must acknowledge the need for source control and dilution of pollutants. So in most cases, we will want a compromise or multifaceted approach—a tight building envelope, increased ventilation, better filtration, perhaps an energy recovery ventilator, etc. For a closer look at ventilation as it applies to indoor air quality, even though it deals with residential systems, Bower's book is interesting and informative. Shirley Hansen's *Managing Indoor Air Quality*, published by The Fairmont Press of Atlanta, GA, is also very helpful.

On the other hand, many critics say runaway ventilation rates are counterproductive and needlessly expensive, including Steve Taylor, a former chairman of the committee for the new standard 62, who said the extravagances of 62-1989 sparked his interest in seeing a new standard written to take its place.

A Health Standard?

There are some who believe that Standard 62R should indeed be a health standard. But to do so would be going beyond the scope of what an engineering society can reasonably expect to produce. These are not, after all, physicians or medical researchers. Also, how do we go about enforcing proper ventilation operation and maintenance of existing buildings—many of which were built even prior to the 1973 standard? It is easy to see how controversy can erupt. To boot, there are volatile changes in some of the new standard's wording—in many places, "it is suggested" will give way to "you must." Imagine the impact of such changes in a court of law in hearing a lawsuit on building design.

NOT an IAQ Standard

As a final note, we must qualify standard 62-2001 and any of its updates: NO standard for ventilation can truly be a standard for IAQ. It is a mistake made again and again. Increasing ventilation does not necessarily remove or eliminate indoor air problems, and should not be perceived as a panacea or "magic bullet." IAQ is multifaceted and problems may be difficult to diagnose and cure.

True, good ventilation or proper ventilation rates are instrumental to ensuring a good supply of indoor air, but it does not necessarily follow that acceptable indoor air quality is always traceable to ventilation systems.

FINDING PRECISE VENTILATION (DEMAND CONTROL)

Demand control is a process in which ambient indoor temperature or build-up of CO_2 levels is measured via monitoring devices, then trigger the ventilation system to bring in more outside air.

The operating principle here is that only enough air will be brought in that is needed for combustion and individual needs, no more, no less. Demand control may be a solution to some building air quality woes if stagnant air is a problem and we want to keep a lid on energy costs at the same time. But the methodology still represents a considerable challenge. If a simple and effective system can be achieved, it could provide a welcome vaccine to many sick buildings.

Using CO_2 Gas as a Trigger

Some facility managers advocate the use of CO_2 gas as an indicator of fresh air demand. If CO_2 were a reliable indicator, this would solve a lot of problems.

But there are challenges to the usefulness of CO_2 as an indicator, and there remain questions as to how much CO_2 should be permitted to concentrate before bringing in fresh air.

The Operating Principle - Because people consume oxygen and emit CO_2, if a person spent an hour in a room with no ventilation, the oxygen level would decrease only 0.5 percent during that period while the CO_2 level would increase 230 percent. That is why it is far easier to measure this increase in CO_2 than it would be to measure the slight fall-off in oxygen levels.

Figure 4-11. A carbon dioxide sensor. When carbon dioxide builds to a trigger point, the ventilations system provides more outside air to the space, a process called demand control. Courtesy: Telaire Systems, Inc.

CO_2 as given off by humans includes a number of by-products with it, such as moisture, odors, bacteria, particulate, even viruses. But how much is too much CO_2?

While there is general agreement that 1,500 ppm or higher borders on the unacceptable in an indoor environment, there is plenty of room here for disagreement among professionals.

Applications - CO_2 sensors are best used to control ventilation rates in non-industrial buildings.

Measurement and control of CO_2 in a building can ensure that outside air is being circulated in the right proportion to the distribution of people within the building. It can avoid costly over-ventilation if the space is unoccupied or under-occupied.

In addition, CO_2 is only intended for indicating or controlling the ventilation rate in occupied spaces. If a space is unoccupied, a CO_2 controller is typically designed to set air intake volume at a

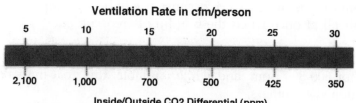

Figure 4-12. Conversion of carbon dioxide inside/outside differential to ventilation rate. Courtesy: Telaire Systems, Inc.

Outside Ventilation Rate (cfm/person)	Indoor/ Outdoor Differential (ppm)	Control Point At Various Outside Concentrations							
		300	350	375	400	425	450	475	500
10	1,060	1,360	1,410	1,435	1,460	1,485	1,510	1,535	1,560
11	964	1,264	1,314	1,339	1,364	1,389	1,414	1,439	1,464
12	883	1,183	1,233	1,258	1,283	1,308	1,333	1,358	1,383
13	815	1,115	1,165	1,190	1,215	1,240	1,265	1,290	1,315
14	757	1,057	1,107	1,132	1,157	1,182	1,207	1,232	1,257
15	707	1,007	1,057	1,082	1,107	1,132	1,157	1,182	1,207
16	663	963	1,013	1,038	1,063	1,088	1,113	1,138	1,163
17	624	924	974	999	1,024	1,049	1,074	1,099	1,124
18	589	889	939	964	989	1,014	1,039	1,064	1,089
19	558	858	908	933	958	983	1,008	1,033	1,058
20	530	830	880	905	930	955	980	1,005	1,030

Figure 4-13. Equilibrium set-points for control of ventilation using carbon dioxide.

minimum setting. It cannot sense the buildup of other contaminants within a space.

Energy Savings - Energy savings can result because ventilation is regulated based on actual occupancy of the space rather than the design occupancy of the space. This can make a lot of sense and tends to be one of its strongest benefits. This type of control is usually applied in conjunction with a minimum ventilation rate or purge strategy to ensure that other building related contaminants do not build up during unoccupied periods. We do not want some level of contaminant building up merely because the building is unoccupied, because it will be too difficult to clear the air all at once when the building is reoccupied.

Acceptable CO_2 Levels - Some researchers suggest a CO_2 level of 600-800 ppm, and while laudable, this goal may also be questioned by others.

Of course, we must compare the inside rate with what we have outside (normally around 350-425 ppm). A level of 500 ppm indoors above an outdoor level of 400 ppm would be logical, not cause for alarm. Even a level of 700 ppm above the outside level translates into an outside air ventilation rate of 15 cfm, which would be well within the realm of providing good air exchange, while 500 ppm CO_2 is equivalent to a ventilation rate of about 20 cfm per person.

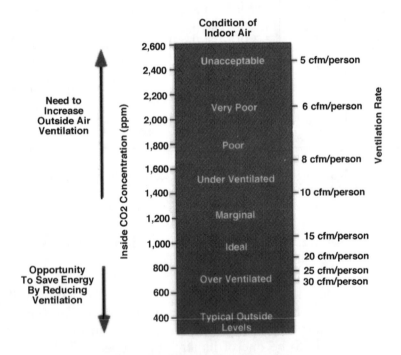

Figure 4-14. Relationship between carbon dioxide levels in a space and ventilation. The point of carbon dioxide-based demand control, as seen in the Figure, is that more precisely meeting the ventilation needs of the space can reduce energy costs. Courtesy: Telaire Systems, Inc.

On the other hand, CO_2 can be used as a surrogate indicator for indoor air quality—meaning that an indoor level of 1,000 ppm as measured above an outdoor level of 400 ppm means not enough outdoor air is being brought in to dilute this gas.

What do ASHRAE and OSHA say?

ASHRAE generally considers 1,000 ppm CO_2 above outdoor levels to be an acceptable level, a level that was referenced as far back as the 1929 Building Code, while OSHA has been favoring a level of 800 ppm above outdoor levels. OSHA currently has in place a workplace permitted exposure level which is much higher—10,000 ppm over an eight-hour time weighted average. To encounter such a level would be highly unusual and would almost have to point to some much more serious, unventilated combustion process. We are not talking dangerous levels here—only what is a good range for working day in and day out, for comfort as well as optimum health.

It is clear at this point that no definitive CO_2 level is perfect for all applications. A level of 1,000 ppm CO_2 above outdoor levels should be thought of as no more than a useful rule of thumb at this point in time.

Challenges to Using CO_2 as an Indicator - A study of ventilation rates and CO_2 levels related to a seven-story test building located near St. Louis was reported in an ASHRAE research paper by Andrew K. Persily, Ph.D., and W. Stuart Dois. Daily peak CO_2 levels in this case were in the 500-800 ppm range—as one can expect, higher CO_2 levels were reported when the rate of exchange with outside air was reduced.

The conditioned space consisted of a mostly open area divided by partitions, with some closed-in private offices. The ventilation system consisted of 30 supply air fans. CO_2 was measured against air change rates—this was viewed as another opportunity to investigate the relationship between CO_2 and air changes (in other words, can it really be used as a simple and effective way to control building ventilation?)

At lower ventilation rates, the observed CO_2 levels were indeed higher. However, determining the CO_2 equilibrium concentration point proved difficult—it was never steady enough for a long enough period of time. Instead, it fluctuated quite frequently.

It might, the researchers acknowledged, work better in a building where the occupancy rate does not change as often.

The researchers concluded, "The appropriateness of the assumptions of equilibrium analysis must still be carefully assessed." In short, the usefulness of CO_2 as an indicator of air changes proved inclusive in this case.

Another study of a 22-story office building in Ottawa, Ontario, Canada concluded that CO_2 made a rather poor tracer gas, although the building experienced an average daily CO_2 concentration of approximately 650 ppm. This was in a building with seven CAV HVAC supply system air handlers and two return air handlers. One objective of the test was to evaluate how CO_2-based demand control ventilation would work in this particular building; the other was to evaluate use of occupant-generated CO_2 as a tracer gas to measure the air change rates.

The study was reported in an ASHRAE paper whereby the researchers explored how well this technique might work even though CO_2 "may not be used to measure the total building air change rate because fundamental assumptions are violated, [although] the technique may measure how well the occupants' needs for ventilation are being met."

Results for calculating air change rates in this particular building were not as acceptable as for using another trace gas, in this case hexafluoride.

It should also be noted that in VAV systems, controls play a large, but not always clearly defined, role in regulating outdoor air. The ultimate goal is to maintain a supply flow greater than the return air flow, creating slight positive pressure within the building to keep out contaminants like vehicle exhaust. To do this, we must bring in slightly more air than the building will lose through open doors and envelope leakage. This amount of leakage is not uniform, of course. More will occur when there are more visitors to the building, or when outside winds or weather conditions are acting to depressurize the building. An inversion layer can cause the opposite to happen, force-feeding air into the building and pressurizing it.

This points to some of the failings of a CO_2-based demand system: CO_2 levels will tend to rise when a building is occupied, but this has little to do with building pressurization or contaminants other than CO_2. For further reading, one method for using several parameters to control a VAV system is discussed in "Outdoor Air Flow Control for VAV Systems," which appeared in the *ASHRAE Journal*. Among other things, the authors point out some of the complexities involved in various climates: "The central system control outdoor air flow represents only the first step in implementing ventilation control."

CO_2 Monitors and Air Quality Sensors

CO_2 sensors should not be confused with air quality (AQ) or volatile organic compound (VOC) sensors. AQ and VOC sensors have generated great interest as a possible approach to controlling building air quality.

AQ sensors and CO_2 sensors measure very different things. In fact, because CO_2 is an inert gas, it is one of the few elements that will not cause an AQ sensor to react. Also, most CO_2-sensing technology is quite stable and is not subject to the short-term, random drift found in AQ sensors. If it is desirable to monitor VOCs as a troubleshooting tool to monitor potential pollutant sources, do so. But use caution before allowing such a sensor to control the building ventilation system.

VOC sensors are also subject to drift, leading to unreliable readings due to changes in temperature, humidity or various gases. AQ sensors can drift as chemical compounds react with the sensor surface. Because of the drift and non specific nature of the sensor output, the use of these sensors for indoor air quality diagnostics and control may be limited. AQ sensors are best applied as an indication of dramatic short term change in contaminant concentrations—cases where unusual sources are suddenly released into the air. Most AQ sensors provide a 1-5 or 1-10 scale of output. It is impossible to determine what a "three" reading means versus a "nine" because of the non-specific nature of the sensor. It can tell us something has changed but it cannot really indicate if the change is significant or quantifiable from an air quality perspective.

Future product development on these sensors will likely focus on trying to make them more consistent and able to be referenced to a known standard of air quality.

Air Quality Sensors - AQ sensors are often also called total

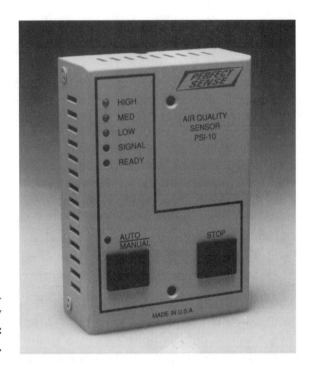

Figure 4-15. A typical air quality sensor. Courtesy: Perfect Sense, Inc.

contaminant sensors because they measure and react to a broad range of diverse compounds in the air. They contain an element which, when exposed to these contaminants, will vary the electrical resistance and produce an output signal, not unlike a smoke detector. The sensor reaction is generally not specific to any one gas. Some consider these sensors useful air quality indicators.

AQ sensors are best used in applications where unusual, non-occupant related sources may periodically surface. As a control, the sensor can activate an alarm or mitigation strategy (activate filters or increase the amount of ventilation by turning on a fan, for example). An AQ sensor, however, cannot indicate ventilation rate. Nor can it necessarily indicate whether safe or harmful concentrations of contaminants are present. It can indicate a general change or trend in the level of contaminants, and this is sometimes useful.

If an AQ sensor is used for demand control, ventilation is regulated based on the actual presence of the pollutants sensed.

Comparison	Broad Band Air Quality Sensors	Carbon Dioxide Sensor
Gases Measured	Wide range of non-specific gases.	Carbon Dioxide
Measurement Units	Cannot be referenced to any known measure but can be individually "tuned" to a building space.	parts per million (ppm)
Measurement Range	Typically 0-5 or 0-10 units	0-5000 ppm
Resolution	1 unit (non quantifiable)	±20 ppm
Common Interference	Temperature & Humidity Also, low levels of silicon vapors and other hydrocarbon species may cause irreversible saturation damage.	None
Calibration	Cannot be calibrated to any referenced standard. Will tend to normalize in the direction of conditions it sees most often over the short term.	Use calibrated gas or recently calibrated instrument as reference.
Type of Technology	Interactive Sensor chemically reacts with pollutants and eventually changes/degrades sensitivity over time.	Non-Interactive Sensor does not come in contact with sampled air. CO_2 concentration determined by infrared light interference of CO_2 gas.
Drift	Constant and unpredictable.	100 ppm per year or less (linear). Can be reduced to 30 ppm annually with self calibrating software.
Best Application	To measure changes in indoor conditions where unusual non-occupant related pollutant sources are a concern. (e.g. printing plants, night cleaners using potent cleaning solutions).	Ventilation control to ensure that target cfm-per-person outside air ventilation rates are maintained at all times in occupied spaces. Energy savings can result where occupancy is variable or intermittent. (e.g. schools, meeting rooms, offices).
Correlation to Ventilation Rate	Levels will probably be higher when there is less outside air ventilation and inside sources of sensor-sensitive pollutants are higher.	The difference between inside and outside concentrations can be directly related to the cfm-per-person of outside air delivered. (e.g. a 700 ppm differential = 15 cfm/person.)
Recognized in the ASHRAE 62-1989 Standard "Ventilation for Acceptable Indoor Air Quality"	No. The "Air Quality Procedure" in the Standard allows for control of ventilation based on reducing concentrations of specific contaminants. Since the sensor does not sense specific gases in known concentrations, it is impossible to know if acceptable contaminant concentrations have been achieved.	The "Ventilation Rate Procedure" in the Standard establishes specific cfm-per-person guidelines for various for applications. Appendix D of the Standard provides the rationale to correlate CO_2 concentrations to specific cfm-per-person ventilation rates.

Figure 4-16. Comparison of carbon dioxide sensors and air quality sensors. Courtesy: Telaire Systems, Inc.

Note that this may or may not conflict with established ventilation codes. Energy is saved when pollutant loads are low and outdoor air can be reduced, usually before or after normal occupancy hours. Where a CO_2 sensor would specifically reduce airflow during unoccupied periods, an AQ sensor may actually maintain ventilation rates during unoccupied periods if there are significant sources still exist within the building.

One problem with AQ sensors is that they often have no way of distinguishing a potentially harmful air contaminant from a harmless trace gas resulting from somebody's perfume or aftershave. Nor is there any widely accepted way to calibrate these sensors to a known measure of contaminant concentration. At best, they might be tuned to the building conditions in which they are operating. The problem with these sensors lies in the broad spectrum of gases they monitor.

Volatile Organic Compound Sensors - VOC sensors detect a wide variety of gases, including CO_2, carbon monoxide, hydrogen, methane and air conditioning refrigerants such as R-123. Gas-specific sensors are available for monitoring other potential pollution sources: CO_2 refrigerant, CO_2, etc. However, we cannot determine from an alarm signal which gas is present or the level of gas in the air. The signal may come from stray vehicle exhaust, from spilled chemicals in a storage room or even because an employee near the sensor is wearing too much perfume. Since the output does not represent a specific level of a known gas, we must attempt to establish a baseline for what it is we are trying to measure. Additionally, if the output rises and stays at an elevated level for a period of time, this new level is automatically established as the new baseline.

One common argument used for VOC sensors is that they require no calibration. On the other hand, a sensor that is not gas-specific and has a non-statistically defined output cannot be said to be out of (or in) calibration.

BACKDRAFT AND CARBON MONOXIDE

If you have ever seen the Ron Howard movie, *Backdraft*, you know what firefighters have long known—that is, how

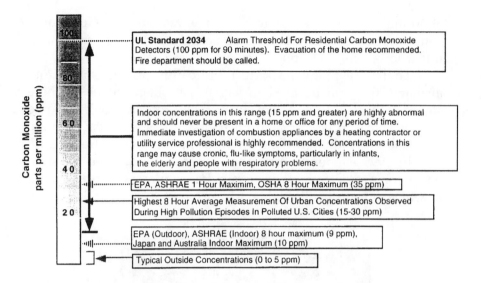

Figure 4-17. Permissible and unsafe exposure levels to carbon monoxide. Courtesy: Telaire Systems, Inc.

powerful a simple thing like imbalances in air pressure or air supply can be. No chapter on ventilation would be complete without a few words on backdrafting. Scary, certainly deadly, and not just in a fire. Every building, commercial or residential, needs some minimal amount of ventilation for combustion and dilution, and it does not necessarily have anything to do with indoor air quality. Backdrafting and leakage from the heating plant continues to be a widespread industry problem.

Much has been done to increase the awareness of this danger. There is still the occasional, unfortunate news story that appears when a faulty furnace or installation kills a family in their sleep. Such tragedies can be averted, however. Similar accidents involving commercial buildings are fewer, but still must be actively guarded against—imagine a deadly carbon monoxide leak in a building that houses dozens, or even hundreds of workers.

Carbon monoxide (CO) differs greatly from carbon dioxide (CO$_2$). CO$_2$ even in small amounts, is a highly toxic poison. It rapidly robs blood of its oxygen-carrying ability. CO$_2$ is a rather benign indicator of poor ventilation, and has to be present in abnormally huge amounts to be dangerous. The effect of a high concentration of CO over a brief period may be the same as a much lower concentration over a longer period.

There are many backdraft indicators and CO monitors available that can be used to signal a problem. Some of these signal with an audible alarm when they detect a problem. Some of them have to be "read" on a regular basis and generally are not practical, although their low cost makes them attractive.

In 1995, Underwriters Laboratories (UL) changed its standards for CO detectors after a number of alarms were sounded in Chicago, causing firefighters to respond to some 1,800 "false

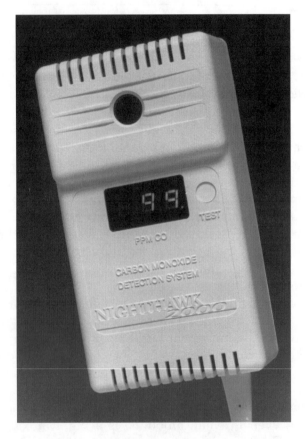

Figure 4-18. A typical carbon monoxide monitor. Courtesy: Nighthawk Industries.

alarm" calls within 24 hours. UL said the detectors may have been overly sensitive and therefore sounded prematurely. At the time, UL also called for changes in the instructions on the packaging and on the detectors themselves to instruct consumers to call a heating and cooling contractor rather than their local fire department if the alarm sounds—unless they are experiencing any symptoms of CO poisoning, in which case an emergency truly exists.

Sources

"The Air We Breathe Indoors," National Institute of Public Health, Sweden, 1995.

Bower, John, *Understanding Ventilation*, The Healthy House Institute, 1995.

Dozier, Jack, "VAV and IAQ," *Engineered Systems*, Business News Publishing Co., November/December 1993.

Eto, J.H., PE and Meyer, C., "The HVAC Costs of Increased Fresh Air Ventilation Rates in Office Buildings," ASHRAE paper, 1988.

Fresh Air Newsletter, The Trane Company, Fall 1994.

Halliwell, Jack and Wheller, Art, "A Preview of The New 62," paper presented at Indoor Environment '96 Conference, Baltimore, 1996.

Hays, Steve M., Gobbell, Ronald V. and Ganick, Nicholas R. *Indoor Air Quality: Solutions And Strategies*, McGraw-Hill, Inc., 1995.

Helm, Randall S. and Bauman, Fred S., PE, "The Healthy Office" series, Steelcase.

Home Answer Book, Hearst Books, 1991, p. 61.

Invironment, The Handbook of Building Management And Indoor Air Quality, Powers Educational Services, Landis & Gyr Powers Inc., 1991.

Kemp, Steven, Kuehn, Thomas H. and Pui, David Y.H., "Filter Collection Efficiency And Growth of Microogranisms on Filters Loaded with Outdoor Air," ASHRAE paper, 1995.

Kirihara, Jay, Editor, 75 *Ways to Improve Your Facility's IAQ*, Milwaukee: Trade Press Publishing Corp.

Meckler, M., PE, "Evaluating IAQ Compliance in Buildings Utilizing Carbon Dioxide Level Monitoring," ASHRAE paper, IAQ '92.

Pedelty, Joe, Holcomb Environmental Services, Olivet, MI. "Ventilation And Indoor Air Quality: Factors That Affect The Results," ASHRAE paper, 1994.

Proclaiming The Truth, ASHRAE Centennial book, Atlanta: ASHRAE.

Ventresca, Joseph, "Improving IAQ with Ventilation Operations And Maintenance," Healthy Buildings. National Coalition on Indoor Air Quality (NCIAQ).

Vergetis Lundin, Barbara L., "Developing an IAQ Action Plan," *Building Operating Management*, February 1996.

Warren Technology, Hialeah, FL, "HVAC Special Report" Vol. 1, No. 3.

Chapter 5

Improving Filtration: Recycling Air

Filtration has come a long way to cure IAQ ills. Like most prescriptive cures, filtration alone will not solve problems such as dirty ductwork or improperly stored chemicals, or by itself kill mold and germs. But new filters and air cleaning devices can help clean up our indoor air and allow us to recycle indoor air without bringing in overly wasteful amounts of outdoor air.

INTRODUCTION

All building HVAC systems share something in common: various mechanical parts with many yards or even miles of twisting ductwork that can harbor dead birds, rodents, rotting leaves, dirt and dust, mold, fungi and construction debris. At least one duct cleaner claims to have uncovered a tradesman's leftover lunch. The first line of defense in preventing these contaminants from accumulating within an HVAC system is through the use of filters.

In the ventilation system, the filter's job is to catch particles. The filter is usually located in the air handler along with coils and equipment for heating, cooling, humidification and dehumidification. Outside air, mixed with return air that is recirculated from indoors, is sent through this "line of defense" to snag airborne particles. Filters carry an efficiency rating—the more efficient, the better the filter is at catching particles.

Figure 5-1. Filtration equipment: ASHRAE-grade air filters, high-efficiency particulate air (HEPA) filters, gas and vapor absorbers, filter system hardware and specialty items. Courtesy: Farr Company.

Traditionally, filtration's main role was not to clean the air for breathing, as one might assume, but mainly to keep "large" particles from being ingested into the mechanical equipment where they could cause damage. In the design of HVAC systems, filtration was an afterthought, and a not very conspicuous one at that. Filters were often replaced on a haphazard basis, almost always with like-for-like products. Most filters were cheap, disposable and easily forgotten.

Today, filtration can play a larger role if we are to clean up our indoor air, and deserves notice. New high-quality filters and air-cleaning devices are available from dozens of manufacturers. The industry is moving toward standardization of efficiencies at catching particles to simplify selection and specification. If the right filter is selected and the filtration system is properly maintained, significant IAQ benefits and energy cost savings can result.

THE FILTRATION STRATEGY

Recycling is a good buzzword for our conservation-conscious society. So why not recycle air? In a manner of speaking, we do it all the time. Air that circulates through our buildings is drawn through filters to be either cooled or heated according to need. A bit of what we inhale in every breath is recirculated air, and there is no harm in that.

Current thinking in ventilation standards calls for drawing in more outside air in an attempt to dilute contaminants and to ensure a constantly renewed supply of fresh air. Most buildings designed between 1973 and 1989 supply 5 cfm per person of outside air—we would have to at least triple that to update the ventilation system to ASHRAE Standard 62 recommendations. This comes with an energy penalty, however, and it is worth mentioning that some outdoor air is not all that "fresh" to begin with. Imagine a hospital emergency room located in the downtown of a major metropolitan area like Los Angeles, for example. Or a manufacturing clean room for sensitive electronics located in downtown Dallas or Detroit.

If highly efficient filters are properly selected, installed and maintained, we can do a better job at cleaning both incoming outdoor and recycled indoor air, reducing the amount of outside air required. Because outside air takes more work to heat and cool, if we increase the ventilation rate using outdoor air only we may have to increase the size of the heating and cooling plant, adding installation and operating costs. A percentage of this capacity and associ-

Figure 5-2. Washable and reusable filter. Courtesy: Newtron Products.

ated costs can be eliminated via relying more on recycled indoor air, which costs less to heat and cool.

Let's look at an example:

To increase the amount of outdoor air brought into its newly expanded convention center for ventilation purposes, Kansas City (MO) planned for a large and costly chiller plant to cool this incoming air. Looking for a way to reduce equipment costs without compromising performance, a means was suggested to cut the total prescribed outdoor air volume demand in half—from 171,500 cfm to 85,750 cfm—thereby saving 400 tons of

Figure 5-3. Grille-mount air filter. Courtesy: Trion.

APPLICATIONS	BENEFITS OF ENHANCED FILTRATION
Protect mechanical equipment such as heat exchange coils	Increased system efficiency from clean components. Lower maintenance costs. Reduced microbial growth.
Protect systems such as air distribution ductwork, outlets, and porous components	Increased system life. Reduced cleaning cost. Increased efficiency. Lower maintenance costs.
Protect occupied space such as wall and ceiling surfaces	Lower maintenance costs on equipment. Lower housekeeping costs.
Protect occupants from contaminant exposure	Increased comfort. Lower health risks and costs. Reduced absenteeism. Heightened productivity.
Protect processes such as pharmaceuticals and electronic chip fabrication	Avoidance of product failure. Reduced downtime.
Provide clean makeup air to use as high quality dilution air	Enhanced energy management. Lower operating costs. Compliance with NAAQS Standard.
Provide source control to avoid reentrainment into the return public air	Reduced cost of containment control. Occupant control of air quality.
Augment ventilation air through treatment of the return air	Increased dilution capacity of ventilation air.
Protect environment from contaminated exhaust	Avoidance of reentrainment. Compliance with clean air regulation.

Figure 5-4. Uses and benefits of filtration.

chiller plant capacity. The solution consisted of replacing "normal" 50 percent efficiency (based on ASHRAE Standard 52.1-1992 dust spot testing) air filters with 95 percent filters incorporated into a multistage filtration system. The manufacturer was The Farr Company of Los Angeles.

By substituting cleaned, filtered, recirculated air for some of the projected outside air demand, the city was able to reduce the amount of outdoor airflow from 15-20 cfm per person to 7.5 cfm per person, saving installation costs and associated operating costs for 400 tons of cooling capacity.

FILTRATION AND ASHRAE 62-1989

Ventilation standards for clean air, including ASHRAE 62-2001 discussed at length in Chapter 4, largely failed to address filtration as an IAQ tool. Standard 62-2001 does not require filtration *per se* for our indoor ventilating needs, but rather suggests the use of filters in situations wherever the need for makeup air is high, or where it is necessary to remove contaminants from recirculated airstreams.

Three Times For Good Luck

The management company at South Coast Plaza, because of poor outside air in smog-shrouded Cosa Mesa, CA, decided to "triple-filter" the air at its four million sq.ft. high-rise office complex. Outside air is filtered once as it comes in through a main duct, then again as it comes from individual fans to each floor. It is filtered a final time when it goes into the conditioned space. Fan power boxes are used to keep the air moving at all times, making the most of the filtered air and keeping dirtier outside air infiltration at a minimum. Stanley Taeger, director for Office Property Management at South Coast, said, "In big-city environments, better outside air is not always the case." The cost of the filter changeouts, at $1,500, is considered just another necessary part of the $1 million annual maintenance budget.

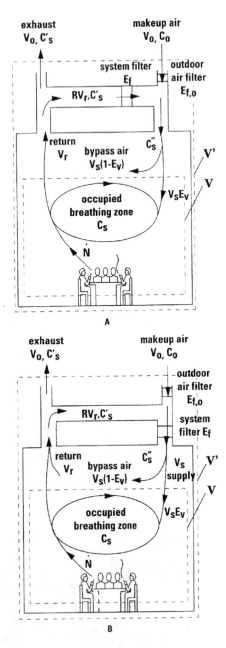

KEY

C_0 - concentration of contaminant in outdoor air intake

C_s - steady state concentration of contaminant in an occupied space

E_f - efficiency of return or mixed air filter

$E_{f,o}$ - efficiency of outdoor air filter

E_v - ventilation effectiveness defined as fraction of supply air delivered to an occupied space

$\dot{N}$ - net internal generation rate of contaminant within an occupied space

R - fraction of supply air which is recirculated

V - occupied breathing zone control volume

V' - total system control volume

V_0 - outdoor air ventilation rate

V_r - return air

V_s - supply air ventilation rate (the sum of recirculation and outdoor air), equals return air

RV_r - equal to recirculation flow

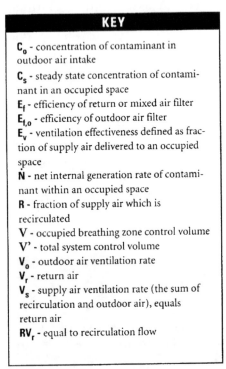

Figure 5-5. Models for contaminant generation and ventilation demonstrating the use of filtration with reduced outdoor air. Courtesy: The Farr Company.

Newer standards attempt to address many areas the older standards did not, including in this case establishment of a minimum filtration efficiency level, filtration as a means of source control, filtration maintenance practices, filtration efficiencies based on particle size, and recommended fractional-size efficiencies.

As currently planned, the new ASHRAE Standard 62 allows up to half of the ventilation air required for a space to be made up of recirculated, filtered air, also stating that ducted supply air systems shall not be operated without filters in place and fitted properly.

Some equipment manufacturers balked at the demand for better filtration. Many types of air conditioning equipment will have trouble physically accommodating what may be larger, bulkier filters, while they also will have to contend with the pressure drop from filters that are denser, and thus inhibit airflow.

There are many new types of filters on the market, however, including some which are impregnated with carbon to remove odors, and some that can be cleaned and reused instead of being tossed away after limited use (see Figure 5-2). The reusable filters could wind up saving money in the long run, but only if the maintenance staff rigorously adheres to the cleaning schedule. An alternative is to hire an HVAC service company that will tend to this task as part of an ongoing service or maintenance agreement.

Just a few of the dozens of manufacturers of high-quality air filtration products include Newtron Products Co. (Cincinnati), Purolator Products Co. (Henderson, NC), The Farr Company (Los Angeles, CA), Airguard Industries (Louisville, KY) and Precisionaire Inc. (St. Petersburg, FL).

FILTER EFFICIENCY

Major changes for rating filter efficiencies have been underway as this book was being developed.

Presently, filters are tested using one of two methods described in ASHRAE Standard 52, "Method of Testing Air-Cleaning Devices Used in General Ventilation for Removing Particulate Matter." The first test method is the weight arrestance test, with a value described as a percentage of efficiency. It is typically used to assess how well low-efficiency filters used in residential applications catch large particles that can damage the mechanical equipment.

It is worth noting at this point that our bodily defense mechanisms are much better able to filter and handle large particles, say five microns and up—including pollen, mold and lint, which are borderline visible to the naked eye—whereas small particles can sneak past and be lung-damaging. Particles under five microns include bacteria, skin flakes and smog.

When we want to assess how well a filter catches smaller particles that are breathable in addition to large particles, the atmospheric dust-spot test is much more helpful. The dust-spot method measures particles in the more reasonable 0.3-6 micron range.

Note that the test method may yield widely disparate results. In other words, if we test a filter according to the weight arrestance method, we may get a wildly different rating than if we test it using the dust-spot method. For an efficiency comparison to have meaning, the filter must be targeted to removing the same size of particles.

Using the dust-spot method, low-efficiency filters are classified as having an efficiency rating of 10-20 percent; medium-efficiency filters have a rating of 30-60 percent, and high-efficiency filters have a rating of 85-95 percent. The rating, as stated, is the efficiency of the filter with respect to the size of particles being trapped.

Note that some filters work better when they are new, while others are more effective when they are partially loaded. Averaging means we may get an efficiency figure that is skewed for the new filter we are installing.

For many buildings, medium-efficiency pleated filters with a dust-spot rating of 30-60 percent are considered a sensible choice. They are fairly efficient, and will provider longer service before clogging than high-efficiency filters. Applications that are to be kept exceptionally clean should contain high-efficiency filters with a dust-spot rating of 85-95 percent.

If we are to rely on filtration as a key element in our IAQ strategy, however, it may be wise to choose the most efficient filters available. For one thing, consider that a fractional efficiency can also be viewed as a fractional *inefficiency*. A filter with a dust-spot rating of 60 percent in an office building, for example, has an *inefficiency* of 40 percent—why let any percentage of respirable, and potentially harmful particles, get through? The facility manager must weigh the priority of proper filtration with the cost and its role in the IAQ program. The additional cost could easily be viewed as an investment if the alternative is increasing the outdoor air ventilation rate.

A key element of filtration, for example, that is tied to efficiency is the amount of energy required to push air through the filter. More-efficient filters tend to hinder airflow, which can raise static pressure, adding to energy costs as well as equipment wear-and-tear. Static pressure is simply the force required to overcome resistance to airflow presented by filters, dampers, etc. A related concept is age of the filter. As a filter becomes dirtier, it becomes more efficient because the dirt particles tend to cling to one another, thus squeezing the air through smaller and smaller openings in the filter, restricting airflow. It is necessary to determine what the needs are and what costs can be justified. The high-efficiency filter will require more frequent replacement.

One option that some manufacturers say is the best way to keep energy costs low and IAQ benefits high in office buildings with VAV HVAC systems is a high-efficiency extended-surface filter with a dust spot rating of 85 percent without a pre-filter.

One thing is certain, and that is office buildings, schools and other applications deserve better. "Unfortunately, the filters in the home and office are nothing more than butterfly nets designed to protect against mechanical wear and tear on the HVAC equipment, similarly to an oil filter in a car," said IAQ consultant Michael A. Price, MAP Environmental. "Today, we must look to filtration to also protect the people within the building, especially on removal of respirable dusts that may be lung damaging."

The cost of high-quality filtration costs less than $0.02 per person per workday, assuming an occupancy ratio of 150 sq.ft. per person ($4.50).

As noted at the start of this section, the criteria for judging filters is changing for how they are specified and applied for different uses. One problem has been that filters are rated using tech-

Apartments	20% to 30%	**Government Buildings**		**Pharmacy**	
Arcades	20% to 30%	Court Room	30% to 35%	Work Room	30% to 35%
Arenas		Legislative Chambers	30% to 35%	Sales Area	30% to 35%
Public Areas	30% to 35%	Committee-Conference	30% to 35%	**Photo Studio**	
Floor	30% to 35%	Offices	30% to 35%	Studio	30% to 35%
Ticket Booth	30% to 35%	Press Room Lounge	20% to 30%	Darkroom	30% to 35%
Stages & Work Rooms	30% to 35%	Toilets	Below 20%	**Pool Hall**	20% to 30%
Bakeries	20% to 30%	Jail Cells	20% to 30%	**Restaurant / Nite-club**	35% to 40%
Ball / Dance Rooms	30% to 35%	Guard Stations	20% to 30%	**School**	
Bank	35% to 40%	Police Station	20% to 30%	Classroom	30% to 35%
Barber Shops	20% to 30%	Fire House	20% to 30%	Laboratories	30% to 60%
Bars and Discos	20% to 30%	**Gymnasiums**		Shops	30% to 35%
Beauty	20% to 30%	Floor	20% to 30%	Music Rooms	30% to 35%
Bowling Centers		Spectator	20% to 30%	Libraries	30% to 35%
Bowling & Spectator	20% to 30%	Locker	Below 20%	Auditoriums	30% to 35%
Cafeterias	35% to 40%	**Hospital**		Gyms	20% to 30%
Church		Patient Room	90% & PF	Locker Rooms	Below 20%
Assembly	30% to 35%	Medical Procedures	90% & PF	Common Rooms	30% to 35%
Classroom	30% to 35%	Operate Delivery	90% & PF	Offices	30% to 35%
Meeting / Activity	30% to 35%	Recovery / Intens Care	90% & PF	Lunch Room	30% to 35%
Coin Laundries	20% to 30%	Autopsy	90% & PF	**Shoe Repair**	20% to 30%
Convenience Centers		Physical Therapy	90% & PF	**Skating Rinks**	20% to 30%
Stores	20% to 30%	Cafeteria	90% & PF	**Small Stores**	20% to 30%
Court Club/Fitness Center		Amphitheater	30% to 35%	**Smoking Room**	80%
Tennis	20% to 30%	**Industrial**	10% to 99.%	**Supermarket**	35% to 40%
Racquetball / Squash	20% to 30%	**Kitchens**	Below 20%	**Swimming Pools**	
Exercise Room	20% to 30%	**Laboratories**	80% to 99%	Pool & Deck	20% to 30%
Department Stores		**Lab Animal Rooms**	Below 20%	Spectator	20% to 30%
Shopping Areas	20% to 30%	**Library or Museum**		**Theater**	
Diners or Fast Food	30% to 35%	General Areas	35% to 60%	Public Areas	30% to 35%
Doctor/Dentist Office		Work Rooms	35% to 60%	Floor	30% to 35%
Waiting	30% to 35%	Rare Books & Exhibits	80% to 99%	Ticket Booth	30% to 35%
Examination	30% to 35%	**Shopping Mall**		Stages & Work Rooms	30% to 35%
Medical Procedure	90% +	Mall	20% to 30%	**Public Toilet**	Below 20%
Dormitories		Store	20% to 30%	**Transportation Terminal**	
Bedrooms	20% to 30%	**Motels**		Public Areas	30% to 35%
Suites	20% to 30%	Bedrooms	20% to 30%	**TV-Radio**	
Lobbies	20% to 30%	Suites	20% to 30%	Booths	35% to 40%
Meeting / Game	20% to 30%	Lobbies	20% to 30%	Studios	35% to 40%
Dry Cleaners		Conference	20% to 30%	Electronic Equipment	80% & PF
Commercial	20% to 30%	Assembly Hall	20% to 30%	Audience	35% to 40%
Coin	20% to 30%	**Nursing Homes**		Stages	35% to 40%
Print / Duplicating	20% to 30%	Patient Room	80%	**Variety Stores**	20% to 30%
Exhibition Hall		Medical Procedures	80%	**Veterinarian**	
Public areas	30% to 35%	Physical Therapy	80%	Kennel	10% to 15%
Floor	30% to 35%	**Office Building**		Operating Room	30% to 35%
Ticket Booth	30% to 35%	Office	35% to 60%	Waiting Room	30% to 35%
Stages & Work Rooms	30% to 35%	Bull Pen	35% to 60%	**Warehouse**	10% to 35%
Florist	20% to 30%	Conference	35% to 60%		
Funeral Parlor	30% to 35%	Reception	35% to 60%		
Garages	Below 20%	**Pet Shops**	20% to 30%		

Figure 5-6. Current suggested ASHRAE dust spot efficiencies for non-smoking applications. For smoking applications, filter efficiency must be 80 percent or higher. Note: PF = "Pre-Filter." Courtesy: ASHRAE.

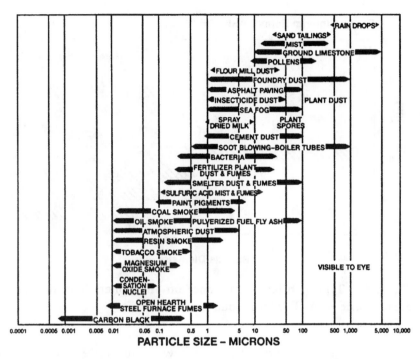

PARTICLE SIZE – MICRONS

Figure 5-7. Particle size (in microns) of typical indoor pollutants. Courtesy: Purolator Products.

nology that is at least 50 years old. ASHRAE has developed Standard 52.2.

According to Barney Burroughs, past ASHRAE president and vice chairman of the ASHRAE Environmental Health Committee, also chairman of the committee for developing Standard 52.2, "Fractional Efficiency of Particulate Filters": "One of the major problems facing the filter specifier and user is the wide range of confusing, exaggerated and/or misleading claims made for filter performance. This is especially true for the range of filters applied in the residential and commercial marketplace." Burroughs said that to evaluate and properly apply filtration equipment, the specifier must consider factors beyond simple efficiency. These include life, capacity, airflow characteristics and cost—including life-cycle cost, which is an expression of the total installed cost plus all other costs assumed in operating the equipment over its projected service life.

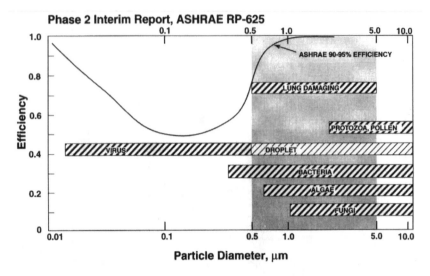

Figure 5-8. Typical filter performance on atmospheric and respirable dust. Courtesy: AAF International.

One major change is that the new testing criteria will use a standard test aerosol, as opposed to simple discoloration of the filter. This will require the use of a particle counter, which is still considered somewhat new technology for this purpose, to report efficiencies in the material being filtered, generally 0.3-10 microns. Not specifying the particle size is detrimental to filter efficiency testing, since some filters will be more or less efficient depending on the size of the particles they are capturing. There is widespread agreement that better filters should be used, that more effective filters will help clean our air—just as there are as many new cases discovered every day of air filters that have gone unchanged for years, have been removed completely from their holders, or have had holes cut out of their centers "to improve airflow."

Hopefully, that is all changing.

FILTER SELECTION

Application and use of something as relatively simple as an air filter can still be somewhat of a mystery, with a lot of unknowns and plain outright mistakes made. Some of this is being cleared up through the efforts of the National Air Filtration Association. Selection of a filter should include criteria that will prevent damage to the mechanical equipment, and allow the filter to catch small particles to enhance IAQ. Besides standard filters, stand-alone and carbon-impregnated filters are available, in addition to filters with microbe-killing coatings and UV light units.

Stand-alone Filters

Filters are not always incorporated into the air supply. There are stand-alone cabinets with fans and filter packs that pull in room air, often using a system of more than one filter including a cheaper, less-efficient "throw-away" pre-filter.

Some come with an electrostatic feature, making them two-stage electrostatic precipitators (see Figure 5-9). The electronic air cleaners incorporate a number of metal plates that carry an electrical charge of about 8,000 volts that draws particulates onto them like a magnet. Drawbacks are a higher initial cost, higher operating cost, maintenance and eventual replacement cost. They also give off minute amounts of ozone, which is another possible contaminant. The plates on some electronic air cleaners may be cleaned in place if they have an automatic wash cycle; some are served by pre-filters or after-filters that must also be occasionally replaced or cleaned. Such devices are usually selected on the basis of 80 percent dust spot efficiency or higher for commercial use.

Maryland Schools Clean Up Their Air

Schools represent an IAQ problem that directly affects 51 million children and school employees. The State of Maryland has an impressive IAQ program that dates back to 1986 and involves all 24 of the state's school districts. Efforts have included the use of better air filters, changing the thinking from picking filters based on the weight arrestance method (large particles) to the dust-spot method (small particles and up). The routine policy on carpeting is to air it out 24 hours prior to installation to reduce the concentration of volatile organic compounds off-gassed, then for 72 hours more before occupancy.

Figure 5-9. A stand-alone electronic air cleaner. Courtesy: Trion Inc.

Carbon Filters

"Partial bypass" carbon filters and carbon-impregnated filters are available that reduce volatile organic compounds in office buildings.

Activated carbon, sometimes used interchangeably as activated charcoal, is a universal absorbent, commonly produced from coal or coconut shells. It is used everywhere from respirators to cigarette filters, as well as in many industrial applications. Carbon can remove gases and vapors that the best filters cannot remove. Absorption is physical attraction to the surface of a solid, almost like a magnet. These particles are often 0.01 microns and smaller, compared to 0.3 microns for the best high-efficiency particulate air (HEPA) filtration. Remember that one micron equals 1/25,400 of an inch. True HEPA filters are the best air cleaners, with efficiency ratings of close to 100 percent (99.97 percent). That is why one should be careful when a filter advertises that it captures up to 90 percent of airborne particles. For one thing, 90 percent of what size particles? Few filters, for example, will be effective at filtering tobacco smoke, especially the tars and nicotine in the 0.01 micron range.

With HEPA filtration, the trade-off is that it takes an abnormally powerful blower or fan motor to draw or push air through the filter, with an associated higher level of energy cost. In addition, beware of the term "HEPA-type" or "HEPA-like." Since they are not true HEPA filters, they may offer inferior efficiencies of only 10-35 percent.

Beware of cheap carbon filters that may have carbon flaking off the filters in the form of dust, adding to the IAQ problem instead of solving it.

Carbon filters often are teamed with a medium-efficiency filter to extend the life of the absorbent. Of course, with any high-efficiency filter, we have to be careful we do not overwhelm the system. As stated, a thicker filter or more dense filter media will screen out more dirt and particulates, but will also slow down airflow and require more work (electrical energy) to push the air through. This is especially true as the filter gets dirty and starts to clog.

Microbe Killers

Some companies, including Aegis Environmental and AAF International, offer filters that feature a coating of an anti-microbial agent. As filters become loaded (dirty) operating in a humid environment, the filters themselves can become breeding grounds for bacteria. Smart maintenance practices will also help solve this potential problem.

Portable Office Air Cleaners

Some companies, such as 3M Company, Trion and Honeywell, manufacture small portable air cleaners designed for individual office use.

UV Light Units

Ultraviolet light is a common tool now used to kill microbes. This has been used before in many hospitals and healthcare situations, but it is finding its way into other applications as well. Some new combinations of filters incorporating UV lights are

FILTER DESCRIPTION	MAX FACE VELOCITY fpm	MAX PRESSURE DROP in H₂O		EFFI-CIENCY[1] %	ARRES-TANCE[2] %	COMMENTS Application & Limitations
		Clean	Dirty			
A Metal Mesh, 2" Deep, Oiled (Cleanable)	520	.35	.5	15-20	50%	Reasonably Effective On Large Particles & Fibers Such As Lint. Generally, Ineffective On Pollen, Smoke & Settling Dust.
B Microfine Glass Fiber Media With Wire Support Grid, 2" Deep (Disposable)	500	.30	.90	30-35	90%	Central Domestic Heating & Cooling Systems. Prefilter For Higher Efficiency Filters. Limited Effect On Smoke & Soiling Particles.
C	500	.35	1.5	50-55	95%	Commercial Fresh & Recirculated Systems. Effective On Pollen & Fine Dust Particles. Limited Effect On Fume & Smoke. Ineffective On Tobacco Smoke.
D High Density Microfine Glass Fibers With Wire Support Grid & Contour Stabilizers, 12" Deep (Disposable)	500	.5	1.5	80-85	99%	Commercial Fresh & Recirculated Systems. Effective On All Pollen & Dust, Most Soiling Particles, Fume Coal & Oil Smoke.
E	500	.65	15	90-95	100%	Pharmaceutical, & Clean Hospital Areas, Effective On Soiling Particles, Coal & Oil Smoke, & Bacteria. Effective Against Tobacco Smoke.
F Continuous Sheet Of Glass Micro Fiber Paper With Corrugated Separators. Wood Board Frame With Gaskets. 12" Deep (Disposable)	275	1.0	3-4.0	99.97[3]	100%	Excellent Protection Against All Smokes & Fumes. Bacteria & All Forms Of Dusts. Surgery Rooms, Intensive Care Wards, Clean Rooms & Pharmaceutical Packaging.

NOTES:

1. Efficiency based on ASHRAE Dust Spot Efficiency Test Method using atmospheric dust.
2. Arrestance based on ASHRAE Weight Arrestance Test Method using synthetic dust.
3. Based on DOP Test (0.3 micron smoke).

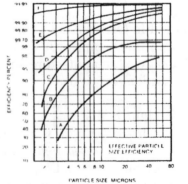

Figure 5-10. Filter selection guide published by a manufacturer. Graphic performance data not shown. Courtesy: Munters Corporation, Cargocaire Division.

showing up for use in HVAC system ductwork. Slime, mold and fungus are simply more likely to be found in dark areas than in direct sunlight. Units are manufactured by companies such as Specialized Ventilation Systems.

For hospital, school, airline and other enclosed public-space applications, special units can be employed that are self-contained. One series of units, called Viotec, is manufactured by UV Technologies of Greenwich, CT. UV light is used around the clock to kill airborne bacteria such as tuberculosis, Legionnaires' disease, infectious jaundice, even *Ebola*. Various models, tested at the University of Cincinnati, have been shown to be 96-99 percent effective at inactivating bacteria, reducing risk of infection. The unit draws room air into one end of the device, draws it through dust filters and past a series of high-intensity UV light tubes, then releases it back into the space out the other end. No UV radiation is released into the space due to the enclosed design.

FILTER MAINTENANCE

Scrupulous planned maintenance is a problem-solver in many IAQ situations, and filtration is no exception. Without

Figure 5-11. Medium- and high-efficiency filters that utilize a replacement cartridge; also shown is a charcoal filter that can remove particles and unwanted odors. Courtesy: Malone & Associates.

proper maintenance, the filtration can actually become a powerful IAQ problem of its own rather than an aid. Poor maintenance can also lead to costly damage of the HVAC system itself.

General Maintenance Checklist

Here is a checklist to help build a successful filtration maintenance program:

1. Check the filter's holding frame for leaks. Air follows the path of least resistance. Small uncaulked cracks may appear negligible, but combined they can add up to a substantial opening. We cannot allow unfiltered air to enter the HVAC system, not only because it impacts on air quality, but because large particles can enter the mechanical equipment and damage it.

2. Check to ensure the outside air louvers are working properly. Many sick building syndrome occurrences are caused by too little outside air. Sometimes, the outside air louvers are restricted below their original design specifications to conserve energy.

3. Check condensation pans for algae build-up that can block drains.

4. Use the correct filters for the area which they are intended to protect. A minimum of 85 percent dust spot efficiency should be used in office areas to reduce housekeeping and provide adequate filtration for the removal of airborne contaminants.

5. Change the filters regularly. Filters capture dust, which is the carrier for bacteria and other microbials. Loaded filters in a humid environment can actually serve quite well as nesting material for microbial growth, then allowing them to be released into the airstream as they develop and grow, a virtual "microbe farm."

    Clean filters also are less restrictive to the flow of air, saving energy. When changing filters, the maintenance personnel should wear proper personal protective equipment, including a face mask, and turn the blower fan off before proceeding.

    Note that low-efficiency filters, if excessively loaded, can become deformed and may "blow out" of the rack, allowing unfiltered air to enter the HVAC system.

6.   Maintenance duties are much easier to perform if the air handler is easily accessible via quick release and hinged access doors.

Inspection Tips

Look on the suction side of the fan in the air handling unit to find the primary filter, which is usually the only filter bank. An inspection of the filter is to ensure there are no leakage points that can serve as pathways for unfiltered air to enter the HVAC system. Inspect to ensure:

1.   That there are gaskets on hinged doors.

2.   That all duct joints are taped or caulked.

3.   There are no leaks around pipes.

4.   That friction-fitting access doors are taped.

5.   There is a trap connected to the condensate drain designed to prevent water from being pulled out by the suction caused by the fan. A dry trap can be a pathway for unfiltered air.

Determining When to Replace The Filter

Here are some rules to help guide the facility manager in determining when to change the filter:

1.   When the air-leaving side of a flat panel filter, such as a roll filter, shows dirt, it should be changed.

2.   Change the filters on a fixed schedule. The typical changeout point for a flat panel filter is 30-60 days. The typical changeout cycle for a two-inch pleated filter is every 3-6 months. By basing filter replacement on a fixed schedule, we may not gain the maximum life of the filter, but we can simplify labor and materials budgeting.

3.   Change the filter just before extremely hot or cold periods, as these are times when the HVAC system works hardest.

4.   Check the pressure reading on the gauge. Note the clean

Filter Selection - Example

The total free air area (FAA) of a filter is dependent on system design and fan selection, and is the function of the HVAC system designer. Determining how much total FAA is required for a system is somewhat complicated. The example below may help, or one can contact the National Air Filtration Association for specific advice: Call (202) 628-5328 or fax (202) 638-4833.

Fan rated to handle 0.8-in. water gauge (WG) resistance in the supply air duct and 0.05-in. WG resistance in the return air duct.

Total external static pressure (ESP) is therefore 0.85-in. WG. In this case, total resistance allowable for duct losses and an air filter that is ready for changeout cannot exceed 0.85-in. WG. Assume a cooling load is 5 tons and with 2,000 cfm system capacity. We will assume ductwork resistance of 0.59-in. WG. That leaves 0.26-in. WG of available static for the dirty air filter.

The final (dirty) resistance of an air filter is generally double its initial (clean) resistance, so we need to go through the steps necessary to select a filter with an initial resistance of 0.13-in. WG (0.26-in. WG ÷ 2 = 0.13-in. WG) when operated at the system capacity of 2,000 cfm.

A filter with a peak arrestance of 79 percent will allow dirt to visibly accumulate on the fan wheel and in the supply air ducts. This figure of 79 percent peak arrestance sounds impressive, but it is considered a coarse filter and would only stop some large particles such as carpet fibers, lint and pollen. It is only marginally effective on particles smaller than 10 microns such as mold, spores and humidifier dust. A micron is one millionth of a meter, or 0.00004 of an inch—the period at the end of this sentence would fit about 400 microns. The tars and nicotine found in cigarette smoke range from 0.01-1 micron, while skin flakes range from 1-10 microns. On the other hand, a 25-30 percent ASHRAE dust spot efficiency rated filter is generally the minimum recommended for use by IAQ experts. Pleated filters in the $7-$14 price range are sold just about everywhere and many meet this 25-30 percent dust spot requirement.

A typical 20 × 20 × 1 pleated filter rated at 0.16-in. WG when operated at 1,010 cfm. Two 20 × 20 filters would be adequate for a

2,000 cfm system but the 0.16-in. WG is higher than the 0.13-in. WG we are seeking. To get the resistance down, we will need to make the filter bank larger than 20 × 40.

The 20 × 20 × 1 filter has 2.78 sq.ft. of FAA (20 × 20 ÷ 144 = 2.78). Divide the FAA into the rated cfm to find the face velocity (1,010 ÷ 2.78 = 363 ft./minute) To determine what face velocity will give us a 0.13-in. WG resistance, divide the desired resistance by the rated resistance (0.13 ÷ 0.16 = 0.813), then multiply the rated face velocity by the result (363 × 0.813 = 288 ft./minute).

To use this particular brand of pleated filter and limit its initial resistance to 0.13-in. WG, the face velocity cannot exceed 288 ft./minute. The FAA of the filter system required for this application is determined by dividing the face velocity into the system cfm: 2,000 cfm ÷ 288 face velocity = 6.77 sq.ft. We will need to select filters from this supplier with a minimum of 6.77 FAA.

These are the FAA of the most popular size filters:

16 × 20 = 2.22 FAA	20 × 20 = 2.78 FAA
16 × 25 = 2.78 FAA	20 × 25 = 3.47 FAA

Two 20 × 25 filters have the required minimum 6.77 sq.ft. We will get slightly more than 2,000 cfm with clean filters, but the system will not drop below 2,000 cfm if the filter gets changed when resistance has doubled. Another direction to use in approaching this problem is to select an HVAC unit with a higher capacity fan.

There are two ways to choose the combination of HVAC unit and air filters. Select the HVAC unit to accommodate the chosen air filter, or select the air filter to match the HVAC unit.

To select the HVAC unit to accommodate the chosen air filter:

1. Calculate the required cfm for the system. Multiply tons of cooling by 400 to get cfm.

2. Find the static resistance required to move the cfm from Step 1 through the air filter selected for the job. Look for this in the filter manufacturer's literature.

3. Multiply above resistance by two.

4. Calculate duct losses for the air distribution system.

5. Add Step 3 and Step 4 results.

6. Select an HVAC unit that meets the required AC tonnage and the static pressure in Step 5.

To select the air filter to match the HVAC unit:

1. Find the external static resistance available from the HVAC unit selected for the job in the unit manufacturer's literature.
2. Calculate the duct losses for the air distribution system.
3. Subtract Step 2 results from Step 1.
4. Divide Step 3 results by 2.
5. Multiply the tons of cooling for the HVAC unit by 400.
6. Select a filter that has the proper cfm rating and an initial (clean) pressure drop that matches or is below the answer of Step 4.

Air Filtration Installation Tips

1. Look for dirty air bypass
   - Between filter and frame
   - Between filters (side access systems)
   - Between holding frames
   - Ductwork leaks
   - Access door gaskets
2. Replaced damaged filters
   - Tears
   - Cuts
3. Turbulent or directional airflow
   - Correct (vanes, guides)
   - Change air filter type
   - Use sturdy pre-filter
4. Water, snow and fog
   - Avoid
   - Use appropriate pre-filters
   - Use appropriate after-filters (final filters)
5. VAV Systems
   - Use appropriate filters
   - Use support system for bag filters
6. Friction
   - Avoid friction with plenum walls and floor

Courtesy: AAF International (information on selection). The information presented here provides general guidelines only; for specific applications, consult a filtration of indoor air quality expert.

Figure 5-12. The filtration chamber featured removes up to 95 percent of all particles, including microscopic viruses and bacteria that contribute to cold, flu and allergies, according to the manufacturer. Courtesy: Honeywell.

pressure drop and change when that pressure has doubled. A typical flat panel filter should be changed when the gauge pressure reads 0.5 inches. A typical WG two-inch pleated filter should be changed when the gauge pressure reads 0.9 inches.

5.   Pre-filters extend the life of final filters but they also slow down the formation of the filter cake which is necessary to increase the efficiency. The final filter will operate below its average efficiency for a longer time period without the beneficial filter cake. Consider operating clean final filters 30-60 days before installing the pre-filter. This should not be done when the final filter is on the HVAC unit-leaving side.

Sources

Allegrati, Richard; "Air Cleaning And Filtration: The State of The Art," Indoor Environment Conference presentation, Baltimore, MD.

Burroughs, H.E. Barney; "ASHRAE Developing New Testing Method, Rating System for Filters," Indoor Environment Conference, Baltimore, MD.

Burroughs, H.E. Barney; "New Filtration Standards: 52.2 for Particulates And 145P for Gaseous Contaminants," ASHRAE paper, IAQ Conference.

Enviros, The Healthy Building Newsletter. Philadelphia: Department of

Health And Human Services, Region III, Editor Frank A. Lewis.
Farr Co., case history, El Segundo, CA.
Glasfloss Industries product literature, Millersport, OH.
Japsen, Bruce, "The Quest for Quality Air," *Buildings*, March 1995.
Ottney, Thomas C., Ohio Air Filter, product literature, Sylvania, OH.
Price, Michael A., MAP Environmental, Washington, DC.

Chapter 6

Controlling Humidity and Moisture

A proper amount of moisture in the air—not too much, not too little—can be a boon to managing indoor air. This chapter will serve as a guide to what a building's needs are regarding humidification and dehumidification, and what a facility manager should be aiming to avoid problems with air that is either too moist or too dry.

INTRODUCTION

The best filtration system in the world, the cleanest ductwork and the most stringent attention to chemical storage will not help a building that has moisture problems. While the finger of blame for IAQ problems tends to point to ventilation, another major culprit is microbial contamination resulting from high levels of moisture. Moisture can also damage carpeting, wall-covering and other building materials, in addition to being integral to thermal comfort, the focus of many IAQ-related complaints.

We tend to think in terms of temperature when it comes to comfort. It may be true that when the common complaint comes, "It's too hot in here," more heat may be building up in interior spaces that the design capacity of the HVAC system can remove fast enough. Indoor temperatures in this case are too high for optimal comfort. But if the temperature is 70°F, we view this as optimal, although it may definitely not be. We must factor in the effect of humidity, or moisture carried in the air.

There is a joke that goes, "What is the difference between hell and a summer in Michigan [you can substitute your own favorite here] in the summer?" Answer: "In hell, it's a dry heat." We have all heard that cliché, "It's not the heat, it's the humidity."

As relative humidity (RH) increases, our ability to lose heat from our bodies from sweating and evaporation decreases, making us feel as if the area is hotter. Low humidity levels, conversely, require higher temperatures for comfort. Moisture in the air is desirable. The trick is to avoid extremes. Whereas low RH can result in taxing the HVAC system to provide more heat, high RH can result in perceptions of "hot, stuffy air," and can promote mold, mildew, allergies, asthma, respiratory infections and pests.

Excess or unwanted moisture in the environment has also proven harmful to furnishings and even the building envelope, leading to structural damage or premature deterioration.

A middle range of humidity minimizes growth of bacteria, mold, fungus and pests, as well as promoting a more even range of thermal comfort conditions. Our humidity strategy therefore includes finding the desired range, maintaining it via humidification and dehumidification equipment, and locating and removing unwanted sources of moisture.

HUMIDITY BASICS

Humidity is a term used to describe moisture carried in air. Relative humidity (RH) is a measurement of the amount of moisture carried that the air can hold at a specific temperature. If the air in a room holds 60 percent of the moisture it is capable of holding at that temperature, its RH would be at 60 percent. Absolute humidity is presumably uniform throughout a room, while RH varies from room to room and may also vary from space to space. One hundred percent RH means that the air is completely saturated and can hold no more moisture.

The ability of air to hold moisture depends on its temperature. Warm air holds more moisture than cool air. As air cools, it releases its moisture, which is why the highest RH in a room is near the coldest surface where warm air cools (cold windows fogging up in winter are common evidence of this phenomenon).

The point at which moisture condenses out of the air onto a nearby surface, such as a window, is known as its dew point.

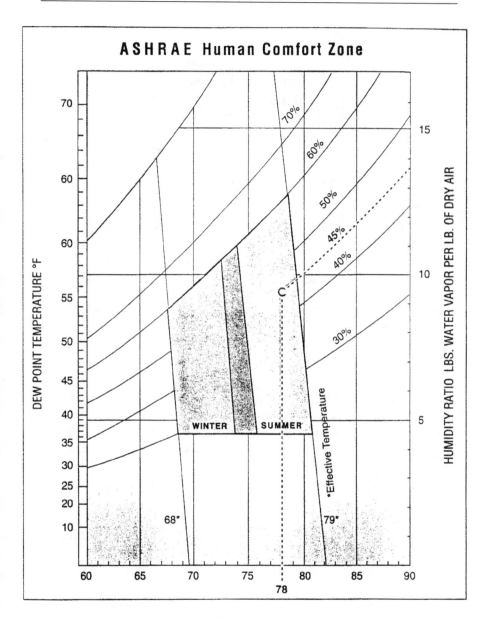

Figure 6-1. Predicted summer and winter human comfort zones, expressed within the context of the relationship between temperature and humidity. Source: ASHRAE *Handbook of Fundamentals.*

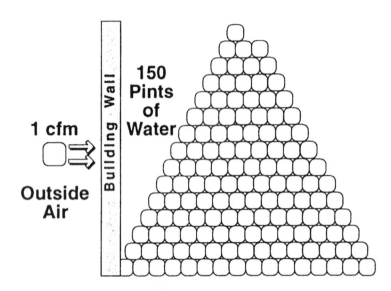

Figure 6-2. Infiltration of just one cubic foot per minute into a building can produce 150 pints of water. Courtesy: CH2M Hill.

HUMIDITY AND MICROBES

An oft-quoted NIOSH study released in 1989 points to inadequate ventilation as the source of 52 percent of all IAQ problems. Microbial contamination lagged at just five percent, with the remainder attributed to other factors that included indoor, outdoor and building material contamination.

Additional studies, however, now report that microbial contamination can account for anywhere from 10-50 percent of building IAQ problems. A paper presented at the May 1993 American Industrial Hygiene Conference analyzed the causes of published IAQ investigations during the 1980s. The analysis showed 15 out of 21 cases of building-related illness to be linked to microorganisms—71 percent.

What does the rise in cases of microbial contamination have to do with humidity? Bacterial growth increases below 30 percent RH and above 60 percent. Viruses proliferate at RH below 50 percent and above 70 percent. Dust mites thrive in humidity levels above 50 percent. Hay fever incidents increase at levels above 60 percent RH and below 40 percent. Controlling humidity in our office buildings, unfortunately, is often still not viewed with the same importance as many other common functions. But controlling humidity is essential to preserving quality indoor air. This may be especially important if we increase ventilation from say, 5 cfm per person to 15 cfm per person, and bring in more humid outdoor air. Some of the worst IAQ problems ever encountered are a direct result of interior humidity gone wild.

Note that humidity conditions and moisture are not solely to blame for microbial growth. Poor ventilation must also be improved to correct problems.

Microbial Sampling: Pros and Cons

A word on microbial sampling: many experts are against it. If fungus or mold are visible on a surface, it is enough to know that it is there, without having to identify the specific variety. Sampling and tests can be expensive as well as inconclusive. There are, however, essentially only a few occasions when microbial sampling might be appropriate, such as:

- To validate a medical hypothesis.
- To provide data for an abatement strategy.
- When confronting litigation or insurance claims.
- To relate to OSHA's "Right To Know" standard.

FINDING THE RIGHT HUMIDITY LEVEL

As stated above, air that is too humid invites microbial growth on interior surfaces. This growth can harbor harmful bacteria, mold and fungus, but it can also lead to deterioration of furnishings and the building itself. Too-humid air or moisture can actually, over time, destroy a building. According to James D. Connolly, who wrote a paper titled, "Humidity and Building Materials" for the Bugs, Mold & Rot II workshop proceedings:

> **Water in its various forms does more damage to building materials than any other single antagonist. Even materials within a building envelope are vulnerable to attack from water, and some materials are particularly vulnerable. To minimize deterioration, the humidity in a building envelope must be properly controlled.**

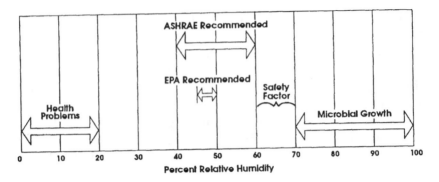

Figure 6-3. ASHRAE- and EPA-prescribed levels of relative humidity. Courtesy: CH2M Hill.

Too-dry air, on the other hand, will make what should be an adequate temperature feel too cool. Too-dry air makes our throats tighten up, our nasal passages constrict, our eyes irritated—conditions made even worse for contact lens wearers or when combined with tobacco smoke or allergies. Air that is too dry can also damage sensitive electronic equipment by producing static electricity. Air in the average office building without a humidifier, in winter heated with forced air, is typically around 13 percent RH. Compare that to RH in Death Valley of 23 percent or 25 percent in the Sahara Desert.

Generally speaking, during the heating season RH should be above 20 or 30 percent, while during the cooling season RH should be below 60 percent. ASHRAE Standard 62, while not directly referring to humidity, does reference ASHRAE Standard 55 which calls for humidification to meet a prescribed range. According to ASHRAE, 30-60 percent RH is a comfortable range, with 40-60 percent RH an optimum range and 50 percent RH considered ideal. At 50 percent RH, there is a substantial and beneficial relative decrease in the effect of ozone production, mites, fungi, bacteria, viruses, respiratory infections, allergies, asthma and chemical interactions.

CONTROLLING HUMIDITY

Many buildings will need to be both humidified and dehumidified at various times of the year, based on changes of the season and other factors, to prevent indoor air that is too humid or too dry. Many climates experience muggy summers and dry winters. Certain spaces within the building, such as computer rooms, may need to be strictly controlled. Supplemental humidification and dehumidification equipment and methods helping the HVAC system are often the answer.

Another integral component of controlling humidity will be to locate and remove unwanted sources of moisture in the building, such as roof or plumbing leaks. We will discuss strategies for dealing with excessive moisture that may be contributing to high levels of humidity in the next section.

Table 6-1. Acceptable ranges of temperature and relative humidity during summer and winter (applies for people wearing typical summer and winter clothing, at mainly sedentary activity). Source: U.S. EPA, adapted from ASHRAE Standard 55, Thermal Environmental Conditions for Human Occupancy.

Relative Humidity (RH)	Summer Temperature	Winter Temperature
30%	68.5°F-76.0°F	74.0°F-80.0°F
40%	68.5°F-75.5°F	73.5°F-79.5°F
50%	68.5°F-74.5°F	73.0°F-79.0°F
60%	68.0°F-74.0°F	72.5°F-78.0°F

Dehumidification

Dehumidification most often needs to take place during the summer, in the cooling season, when the air is warmer and carries moisture into the building. The basic air conditioning system should account for some of the dehumidification needed. In an air system, a cooling coil in the air handler performs this function. As

Proper Humidity and Human Health

Many articles have appeared over the years in medical publications citing the benefits of proper humidification. A sampling taken from literature for humidification equipment tells us:

"One of the most effective ways to avoid sore throats is to properly regulate humidity, especially during the artificial heat season with its distressing drying of the air in living quarters, particularly sleeping quarters ... It is important to prevent an overly dry environment because it makes people more susceptible to infection."
—The American Academy of Otolaryngology
Head and Neck Surgery Inc.

"There is little doubt as to the value of proper indoor winter humidity to health and comfort."
—The Archives of Otolaryngology

"The proper humidification of the inspired air is extremely important to the proper function of the lungs. There is as yet no cure for the common cold. The most important preventative measure would appear to be the proper regulation of the humidity, especially during the heating season with its distressing drying of the air in living quarters."
—*New York State Journal of Medicine*

"If your child is susceptible to respiratory illnesses during the cold weather you must work at this job of allergy control and humidity control constantly to attempt to prevent these diseases. Your child should be able to stand up against the exposure of bacteria and viruses from other children in the school room if the respiratory membranes maintain their normal strength."
—*A Basic Guide for Parents* by Charles S. Sale, MD.

"The mucus secretion of the nasal glands consists of 96 percent water, but it is more viscous than the mucous secretion of any other part of the body. Even slight drying increases the viscosity of the nasal mucus so that ciliary action is impaired."
—The Archives of Environmental Health

air enters the air handler from the mixed-air plenum, the coil pulls moisture from the air. The condensed moisture collects in a drip pan underneath, where it is drained through a trap. If there are overly humid conditions, however, the air conditioning cannot keep up for a simple reason—it was not designed to cope with the conditions.

Generally speaking, the air conditioning process combines sensible cooling and latent cooling in a 70:30 ratio. Sensible cooling, which is 70 percent of the air conditioning system's function, is direct cooling of the designated rooms or zones. Latent cooling, which represents 30 percent of the air conditioning process, is water vapor removal. Seventy percent of the air conditioning system's primary task, therefore, is direct cooling, while removing moisture is secondary. At partial-load operating conditions which occur for a majority of system operating hours, many commercial buildings require higher ratios of latent cooling up to 40, 50 and even 60 percent. Some buildings will require latent cooling only, without sensible cooling.

ASHRAE Standard 62, while calling for increased outside air via increased ventilation, may compound the humidity problem as most existing facilities do not have the capability of handling the sensible and latent loads required. In new facilities, the cooling capacity is more likely to be present to remove the moisture.

One equipment option that can be considered is the installation of a desiccant wheel dehumidification system.

Desiccant Dehumidifiers - Desiccant dehumidifiers can be added to air conditioning systems that use cooling coils to remove water vapor, particularly useful when moisture loads are high compared to sensible heat loads, or when they peak at different periods (see Figure 6-4). An outdoor air temperature of 65° may be fine for comfort conditions, but it may carry too much moisture unless it is first dehumidified.

In a desiccant system, supply air and even return air (if desired) is passed over a material such as lithium chloride, silica gels or molecular sieves. The water in the material has a negative vapor pressure compared to the airflow moving over it. This draws moisture into the material, drying the air. Many materials have

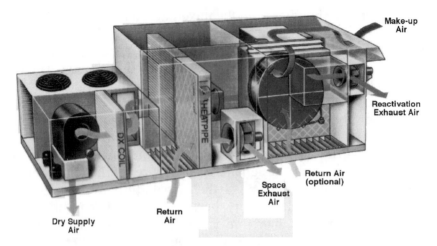

Figure 6-4. A desiccant dehumidification system. Courtesy: Munters Corporation, Cargocaire Division.

capacity to hold moisture, but the materials used in desiccant systems can hold 10-10,000+ percent of their dry weight in water vapor, making them ideally suited for their purpose.

Once the material is saturated with water, a wheel moves new dry material into the path of the supply air, while the saturated material rotates into the path of exhaust air leaving the building, drying the material and taking the moisture outside (Figure 6-5).

In the winter, the system reverses its function, adding moisture to incoming cool, dry air.

Desiccant systems are most popular in retail stores and fast food chains, where condensation builds up on large glass surfaces. Multiple configurations are available based on need (see Figure 6-5 for one manufacturer's list of alternatives).

Besides removal or addition of moisture to incoming air, desiccant systems can reduce energy costs, as was shown at Johns Hopkins University School of Medicine, which experienced a 50 percent reduction in cooling requirements and a nearly 67 percent reduction in humidification and heating requirements.

Figure 6-6 shows a thought process for assessing current dehumidification needs and deciding where and how to use a desiccant system.

Hillsborough County Schools

An interesting case of moisture problems involved two large high schools in Hillsborough County, FL. Each was an eight-year-old design that underwent extensive HVAC upgrades to resolve IAQ problems. Both used a central chilled water plant with two centrifugal chillers, primary/secondary chilled water loops with variable speed pumping, and central station air handlers with variable inlet vanes for VAV terminals. Both were designed under the "old" ventilation standards calling for minimum 5 cfm per person. Space temperatures of 72°F were maintained, but this gave little opportunity for moisture removal or air circulation. Humidity climbed to in excess of 70 percent. CO_2 levels were high, at 2,000 ppm, with significant mold and mildew growth, stale and musty odors, and respiratory ailments among building occupants. After investigation, a target level of 75°F/45 percent RH and 800 ppm CO_2 was established.

Cooling alone would not achieve this. A stand-alone dehumidification unit from CTSI Industries, Longwood, FL, was purchased.

With this, air was initially cooled in a chilled water coil to a dew point of 55°F using 47°F chilled water from the existing chiller system, then further cooled and passively reheated using a small hermetic compressor to conditions of 77°F dry bulb, 57.4°F wet bulb, 28 percent RH.

The dehumidifiers supply dehumidified, room-temperature air to the return side of the existing central station air handlers. Supply air from the central station air handlers was reset to 63°F, dropping RH to below 70 percent, with chilled water supply reset to 47°F to save energy. The existing air handlers were oversized enough to allow a reduction in the chilled water flow to the units, and smaller chilled water control valves were installed. Total outside air was increased from 20,000 to 90,000 cfm, with minimum set points on the VAV system adjusted to provide at least 20 cfm per person at all times.

To moderate an over-cooled space, the chilled water valve reduces water flow to increase the supply air temperature. This affects temperature only. It is independent of ventilation or humidity control, which is important. The combination of equipment and controls wound up with CO_2 levels of 400-800 ppm, temperatures of 74°F-76°F, and RH between 40-45 percent.

Humidification

During the winter, in the heating season, and sometimes in a dry climate, our objective may be to add moisture to indoor air. In office buildings, sufficient moisture almost always has to be supplied via humidification, particularly interior zones that are cooled during the winter using an economizer. One layperson's test of a dry environment is if workers are regularly getting static-electricity shocks after walking on the carpet and then touching a metal object. Employees will provide other clues, such as the presence of portable humidifiers in offices.

Humidifying equipment and methods can be employed to add moisture to the air. Typical humidifiers include direct steam injectors, dry steam and electric. In an air system, the humidifier is located in the air handler. It injects steam into the airflow coming from the mixed-

Figure 6-5. How a desiccant system works. Humid supply air passes over a dry material, which sucks in moisture, drying the air (top). When the material is saturated, it is placed in the exhaust air stream, where it is heated and where it releases the moisture into the exhaust air (middle). The material is then cooled to repeat the cycle (bottom). Courtesy: Munters Corporation, Cargocaire Division.

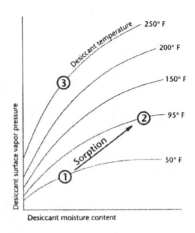

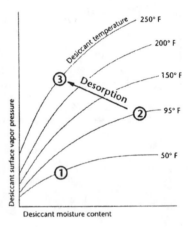

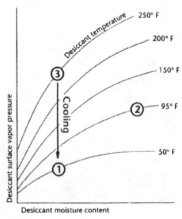

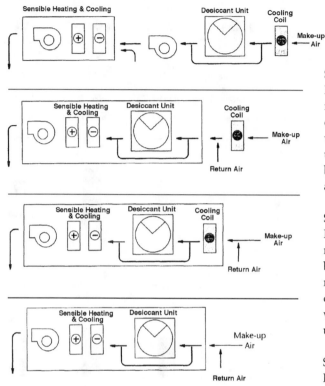

System #1
Pre-cool the make-up air and dehumidify it with a desiccant unit before the air blends with return air from the room.

System #2
Pre-cool the make-up air, then blend it with the return air before dehumidifying with a desiccant unit.

System #3
Pre-cool the blended make-up air and return air before dehumidifying. Cool air improves the performance of the dehumidifier, increasing the capacity of the system, and in some cases allows the designer to use a smaller unit.

System #4
Blend the make-up air and return air before dehumidifying it with a desiccant unit—no pre-cooling. If the make-up air does not carry a significant proportion of the moisture load, it may not be useful to pre-cool in front of the desiccant unit.

Figure 6-6. Comparing desiccant system options. First, four options are identified (above), then compared on the following pages. Illustration, including text (adapted for use here), is courtesy of Munters Corporation, Cargocaire Division.

	System 1	System 2	System 3	System 4
Cooling system size (tons)	Inadequate	17.2	21.8	16.9
Desiccant dehumidifier size (sq.ft.)	Inadequate	3.75	7.5	7.5
Desiccant reactivation (Btu/h)	Inadequate	83,700	109,145	127,121
Reserve moisture removal capacity (lb/hr , % of load)	Inadequate	3.4 (13 %)	41.8 (161%)	28.8 (101%)

Figure 6-6 (continued, including the facing page right). Courtesy: Munters Corporation, Cargocaire Division.

System Comparison

The diagram on the facing page (left) shows a logical thought-path to the dehumidification system that will cost the least to install and operate. But note that there may be many circumstances that alter the general case. It is essential that the dehumidification system designer understand the basic purpose of the project. This understanding provides a map through the maze of trade-offs between the capacity, energy and first cost of dehumidification systems.

System #1: This system has enough capacity to remove the load from the make-up air, but cannot remove the internal load from the room. If the volume of make-up air were larger, the internal load smaller or the control dew point higher, this system would be the smallest and most economical to operate. However, in this case, such an arrangement cannot work.

System #2: This system is least costly to purchase. It has the smallest desiccant unit, a small cooling plant and the least amount of reactivation energy. However, it just barely removes the load from the room. The designer may want to recheck the moisture load calculations if this system is chosen, or perhaps use the next size desiccant unit.

System #3: This system provides more moisture removal capacity than System #2 by increasing the desiccant dehumidifier size and by improving desiccant performance with cool inlet air at the same time. This system will be the most costly to install and operate. However, if the original load calculations were perhaps more of a guess than a firm calculation, or if there is a need for future expansion, this system may be a wise choice.

System #4: This system is more costly to operate than System #1, but it will certainly be the least expensive to operate. Using waste heat to reactivate the desiccant keeps energy costs to a minimum. The cooling coil only uses the minimum chilled water necessary to remove the sensible heat load—not to pre-cool to improve desiccant performance. On an annual basis, this system will probably cost less than half the operating cost of System #3, but will still have the same reserve capacity if the designer installs a high-temperature reactivation heater and extra cooling for future use.

Note: Any change in these circumstances would change the advantages and disadvantages of these different configurations, as noted above. In particular, increasing the make-up air quantity can have a major effect—making System #1 the best choice to keep initial and operating costs at a minimum.

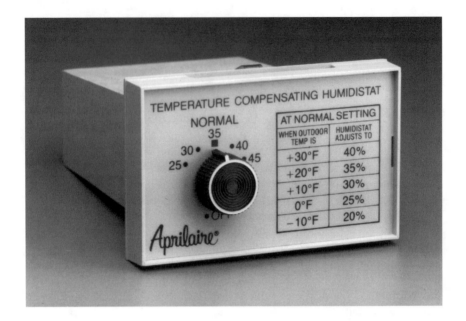

Figure 6-7. A typical humidistat. Courtesy: Research Products.

air plenum to increase its moisture content. Humidified air exits into a duct for distribution. Desiccant wheel dehumidifiers, mentioned above (see Figures 6-4 through 6-7), can also add moisture to incoming cool, dry air during winter.

Note that humidifiers were once suspected of being a rampant source of microbial contamination, especially those that attach to furnaces and deliver moisture to the hot air path through a sponge-type reservoir with rotating drum, flow-through or steam design. As noted in Chapter 2, humidifier fever is becoming a more common malady found in office buildings. Findings of a report done by Pennsylvania State University in 1991, however, gave a clean bill of health to this type of equipment when they are properly installed, operated and scrupulously maintained. Particle sampling in this case did not indicate that the humidifiers were contributing solid particles to the air stream. Nor did particulate matter in the ductwork differ appreciably in any way from that of incoming air.

One key term here is "scrupulously maintained." By fixing one problem, namely air, a facility manager can get in even more trouble if the humidifier is not properly maintained using recommended procedures provided by the manufacturer. Note that occupants sometimes bring in their own portable humidifiers; as these can also be sources of contaminants, they must be properly maintained as well.

Vapor Barriers

Buildings have to be designed from the outset to control humidity and moisture in the form of rain, fog, mist, even snow. Vapor barriers must be installed as the building is being constructed. This can be relatively inexpensive to install, such as plastic sheeting used to form a sandwich layer between an inside poured concrete wall and the outer face brick. But it does require some care to install, so that it effectively keeps out the moisture. On the other hand, these can be expensive and sometimes even impossible to retrofit without major reconstruction.

If a building has been designed with proper vapor barriers in place, there are steps to take to control humidity inside the finished conditioned space, as described in this chapter. In addition, check for irregularities in the wall surrounding the barrier for gaps.

Controls

Typical controls include humidistats, steam control valves and airflow switches (see Figure 6-7).

A humidistat can be used to measure the amount of moisture, tying it to other control measures that can direct airflow or add heating or cooling to curb the problem of too dry or too moist air.

DEHUMIDIFICATION TIPS

Excessive humidity can result from a range of interior and exterior conditions. Like energy, there is also a rule of the conservation of moisture—it can neither be created or destroyed. It simply changes form. It can enter a building as a liquid or as water vapor, and change forms once inside.

If water is left standing in a warm area, it will evaporate into the air. The trick is to control the amount of moisture in the air,

Carpeting Fouls A New York School

Carpeting is a common material that can exacerbate moisture problems. It is not only a good sponge for spills, leaks and tracked-in outside moisture, but serves as a breeding ground for mold once it is wet. Like a glass window or vinyl wallpaper, it can also act as a condensing medium, as was the case in this particular school, as related by Arthur E. Wheeler, PE and Robert B. Olcerst, Ph.D., CIH, OSD in Indoor Air Quality, Professional Reference Program, The MGI Management Institute, produced for ASHRAE:

Subgrade problems are another source that will lead to mold incursions. One school dealt with in New York had a subgrade music room where they had tiers for different sections of the orchestra and it went down subgrade about five feet. The PTA had just gone through and carpeted the music room for the school with indoor/outdoor carpeting; they spent a lot of money. What happened was that the temperature difference was enough to condense room humidity into the carpet; the carpet had a high water content. It bloomed mold and the music teacher herself became hypersensitive. She could not handle sheet music that came out of that music room—within 10 minutes she had bronchial constrictions.

Petri plates placed in the room and in the ductwork and revealed that the concentration was centered on the rug. More colony-forming units per square millimeter of plate were found than the analysts had ever seen before; they said, "This plate is totally uncountable; we've never seen so much mold growth."

The school board, unfortunately, took the ostrich approach; they said that this was her particular medical problem and dismissed the lady. This is not the wisest response because now they had a population of students and teachers who would be going into that room and who were now at risk of being sensitized. What they should have done was to get rid of the carpet and fix the fundamental defect that led to the problem to begin with. The investigators concluded: "These things need to be addressed; you can't shy away from them."

deny it conditions that allow it to condense in undesired areas, prevent leaks and spills, and remove unwanted moisture from the building.

Below is a list of possible causes of high levels of interior humidity, and solutions that can be considered to solve the problem.

Dealing with Indoor Sources of Moisture

Indoor sources of moisture include the HVAC system, plumbing leaks, carpeting on cold floors, house plants, spills, fountains, swimming pools and people:

HVAC System - In the case of the HVAC system, condensation may be forming on poorly insulated piping when a piping system is in operation. The air conditioner units may be improperly sized for the square footage of the building. The evaporative cooling coils in the unit may be dirty, allowing less airflow and causing condensation. Or undesired moisture may be produced during the humidification or dehumidification processes.

The Case Of The Three Schools

Another school-related moisture problem was related in an ASHRAE paper. Three school buildings were studied to examine the effect of lack of humidity control. Many of the occupants of these schools had complained of respiratory illnesses. The authors observed, "The lack of humidity control in each of the three schools contributed to the formation of favorable conditions for visible microbial growth on various surfaces in the buildings. The microbial contamination may have contributed to reported health symptoms in each of the schools."

One school had new carpet adhesive that was unable to cure properly because of the high humidity—as a result, it continued to off-gas volatile organic compounds (VOCs). The authors concluded: "The need for adequate humidity control in school buildings, particularly in hot and humid climates, is reinforced by the conditions in these three schools. The HVAC systems should be designed so that excessive humidity is removed even while meeting the sensible cooling load."

Humidification. After the humidifier adds moisture to the air-flow, the air exits the air handler into a duct for distribution. Moisture often collects in the duct if it makes a turn too close to the humidifier. A useful rule of thumb is to ensure the nearest elbow (turn) is located at least 10 ft. from the humidifier. In addition, if the duct is insulated on the inside with fiberglass lining, the porous lining may be collecting dirt, then moisture, forming an environment conducive to microbials. Or the humidifier may be improperly installed.

Dehumidification. In the air handler in an air system, a cooling coil causes condensation of water into a drip pan, where it is drained into a tap. If the trap was improperly installed, moisture may be collecting only to be evaporated back into the air stream. Or moisture may leak out of an improperly installed trap.

Plumbing - There may be leakage from pipes in the plumbing system. Moisture can also condense wherever cold water pipes meet warmer air, or where warm water is piped through a much cooler area.

Cold Surfaces - Going back to our basic lesson on humidity, RH is highest when warm air carrying moisture is near cold surfaces—car windows fogging up being a typical example. This includes carpeting on cold floors during the heating season. We will discuss this further when we review outdoor sources of moisture in the next section.

Plants - Having many decorative green plants in an office space and overwatering them can produce an abundant source of uncontrolled moisture. Consider that almost all of the water used to water indoor plants enters the air, while only 0.2 percent of this water can be used for plant growth.

There are beautiful hotel lobbies, however, such as the Opryland in Nashville, that are veritable jungles. Our buildings could be stark and lifeless without plants. But if there is already a humidity problem, perhaps artwork, statuary, artificial plants or low-moisture plants would be a better source for aesthetic enhancement.

Spills and Pools - Swimming pools, fountains, atriums, reflecting pools, showers, laundries and kitchens are all areas where water can spill or it is in abundance for ready evaporation into the ambient air.

People - Do not forget that people, too, and their clothes are a source of moisture on the inside of buildings. Every time somebody breathes out, they release moisture-laden air—some more than others, depending on their level of physical activity.

Dealing with Outdoor Sources of Moisture

Moisture can enter a building from the outside from a number of sources, including the HVAC system, leaks, building materials, chinks or cracks in the building envelope or a number of other sources:

HVAC System - Overly humid outside conditions may simply be overwhelming the design capacity of the air conditioning system to dehumidify incoming air during the cooling season.

Roof and Wall Leaks - Water can enter the building through leaks in the roof or the walls, and can be troublesome to locate because the water can travel some distances inside before showing up elsewhere as visible water damage. The water evaporates and promotes higher RH levels in the indoor air.

Building Envelope - Again, using our humidity basics lesson, we know that water condenses when warm air meets a cold surface, which is why we use coasters for cold drinks on a hot day to prevent condensation from damaging a wood table. If the perimeter walls of the building are cold and meet warm indoor air, water will form at corners and at poorly insulated spaces surrounding components of the structural skeleton. Note that water vapor moves through solid materials, even seemingly impermeable marble if it has not been sealed properly, at a rate proportional to the difference in vapor pressures on either side of the material. In other words, the moisture will move faster if conditions on one side of the wall are very dry or very wet, compared to on the other side. Of course, materials vary in their permeance, too. A thin sheet of polyethylene film only 0.006 inches thick will withstand moisture migration much better than an 8-in.-thick concrete block, even though the block is 1,300 times thicker.

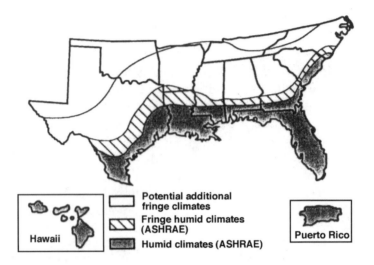

Figure 6-8. Air leaking into the building through gaps in the building envelope is a special problem in regions with hot, humid climates such as the American South. Courtesy: CH2M Hill.

In addition to condensation, chinks in the building envelope may be allowing humid outside air to enter the building and raise interior RH levels. These "chinks" include doorways, windows, cracks and crevices. Air leakage of humid outdoor air into air conditioned building envelopes is being recognized as the major source of most of the IAQ moisture-related problems in the humid South.

Always remember that for humid outside air to enter through the building envelope, three conditions must be present: 1) moisture must be in the air, 2) a hole, opening or pathway must exist and 3) a pressure difference must exist which draws the moisture-laden air into the hole, opening or pathway.

REDUCING MICROBIALS
WITH A HUMIDITY STRATEGY

To reduce microbials that cause IAQ-related illnesses, we know that high humidity conditions can result in microbiological growth in the form of fungus, mold and mildew. Too-low humidity conditions. This can result in thermal discomfort. Our first step, therefore, is to maintain a comfortable RH range of 30-60 percent (or optimum range of 40-60 percent), using supplemental humidifying and dehumidifying equipment as needed depending on changing outside conditions. Other steps include:

1.   Take measurements of indoor and outdoor humidity conditions at different times over the year and in different locations for each measurement.

2.   Check for signs of water damage on carpeting, wall-coverings, drywall and ceiling tiles, then check any water-damaged materials for signs of odors, mold, mildew and pests.

3.   Clean and disinfect any materials affected by mold and mildew using proper safety guidelines and applicable personal protective equipment to protect occupants and the cleanup workers.

4.   Dispose of any contaminated materials such as carpeting and wallboard and replace as needed.

5.   Locate the cause(s) of the damage, which may be internal and/or external.

6.   Correct the problem.

Additional Steps

a.   If there are high humidity conditions, add dehumidification equipment and consider removing plants or replacing with plants which require less water.

b.   Locate sources of high indoor humidity such as kitchens and showers, then isolate the area with dedicated dehumidification equipment or exhaust fans.

c.   Add heat to perimeter areas where condensation forms either by adding heat or ensuring that the perimeters are ventilated with warm air.

Polk County Courthouse

A classic example of a building with a critical moisture problem was the Polk County courthouse in Bartow, FL, near Orlando. In this case, one engineer said that as a building, this courthouse made a good sponge. It wound up costing millions of dollars to remediate (the county later successfully sued the builder and its bonding agent), which included a costly, labor-intensive tear-down of the building's outer brick facade to install a vapor barrier, as well as a complete refurbishing of the interior. This sickest of buildings, which suffered from a number of complaints, traced most of its problems to excess humidity.

Experts said the building soaked up moisture readily from central Florida's warm, semi-tropical climate. In April 1995, a jury awarded the county $26 million plus interest and legal fees against a New York insurance company to pay for the damages, although an appeal was possible as of this writing.

The courthouse reopened in late 1995.

Figure 6-9. Reprinted from January 15, 1995 edition of *The Air Conditioning, Heating And Refrigeration News,* **Business News Publishing Co. Courtesy: Dan Saad.**

d. Add insulation.

e. Locate humidifiers at least 10 ft. from the nearest elbow (turn).

f. Check traps in dehumidification units to ensure they are working properly and not leaking.

g. Check controls for proper operation, including the humidistat, steam control valve and the airflow switch.

SPECIAL PRECAUTIONS IN WORST CASES

Occasionally, a case of fungal growth can be so bad that remediation procedures are akin to the cautions that asbestos removers must use. In such cases, contractors must pay special attention to removing contaminants and protecting uncontaminated areas in the rest of the building from receiving them.

Remember, growth of fungi requires no more than warmth, moisture and some sort of food—unfortunately, conditions which are found all too often in most interior building environments. For food, wallpaper paste, paper, ceiling tile and plain old dirt will do.

In one particular remediation case involving a courthouse building, Clayton Environmental Consultants, Novi, MI, included

---

Wheaton County Courthouse

In the Wheaton County, IL courthouse's Judicial Office Center, a number of problems contributed to conditions that made for a sick building and numerous lawsuits.

A humidifier unit was discovered to be fouled with a buildup of sodium sulfite, which readily converted to a sulfate and found its way into the building's supply air system. In addition, a malfunctioning humidity-level sensor called for the unit to dump additional amounts of moisture into an already too-humid space, forming a visible cloud of moisture that is thought to have reacted with the sulfate in the air to form an acid particulate that was irritating to anybody who breathed it.

the list below in its specifications for the job:

- Isolation of each work area by enclosing it in a double layer of polyethylene plastic sheeting.

- Respiratory and skin protection for workers entering the work area.

- Double bagging of microbially contaminated building materials before removal from the work area

- Use of HEPA vacuums to clean work area surfaces and outer surfaces of bags containing contaminated materials.

- Compliance air sampling for fungal spores both inside and outside the work area following completion of remediation to demonstrate that fungi had been removed and that spores had not been released into occupied areas during remediation.

REPLACING FIBERGLASS DUCT LINING

Earlier in this chapter, we discussed removal of the fiberglass lining in the interior of ductwork as a means to prevent moisture build-up and subsequent microbial contamination in the porous material. Insulation would be wrapped around the outside of the duct and vapor barriers installed.

Before doing so, consider that this may increase transmission of noise from the ductwork, and that there are alternatives available from manufacturers responding to the needs of the market.

Before we review these new products, first be aware that fiberglass is under investigation as a potentially hazardous substance to work directly with. Make sure the contractor who will replace the fiberglass lining is aware of and observes all appropriate and required safety precautions, and provides guidelines and procedures that will ensure occupant safety and health.

The Air Conditioning Contractors of America offers Bulletin #21, Fiberglass Safety as a guide.

Figure 6-10. New interior duct linings are designed with IAQ in mind. Featured is Johns Manville SuperDuct Air Duct Board, incorporating a tough acrylic polymer coating to help control temperature, noise, condensation and microbial contamination. Courtesy: Johns Manville.

Figure 6-11. Tuf-Skin equipment insulation with Permacote coating from Johns Manville contains an EPA-registered anti-microbial agent. Courtesy: Johns Manville.

The makers of fiberglass have improved their products and come out with some innovations such as surface coatings which cut down on noise and microbial contamination. The coating provides a barrier to dust or dirt settling into the wool substrate, and can be more easily cleaned when necessary.

In addition, a number of new products and variants have appeared from insulation manufacturers, including duct liners which carry a tough outer coating to prevent breakdown of the liner, and anti-microbial coatings. The new liners are more easily cleaned than the old ones and stay clean longer because of the slick surface. For the first, time, these surfaces are being designed for easier mechanical cleaning.

An example of a new duct liner product involved a 15-year-old, five-story office building where 20 percent of the occupants reported headaches, sinus infections and respiratory infections. All possible causes were eliminated until it was discovered that the HVAC system had heavy microbial contamination. The duct material was cleaned, but the problems returned. Various coatings were tried, until a coating containing an anti-microbial agent was applied to the internal ductwork surfaces.

TEMPERATURE

While in this chapter we have focused primarily on prescribing and controlling humidity levels, temperature is certainly a primary concern with thermal comfort.

ASHRAE publishes guidelines for temperature ranges, along with relative humidity ranges based on temperature, in its Standard 55, Thermal Environmental Conditions for Human Occupancy. Like the Standard 62's definition of acceptable IAQ, Standard 55-1981 bases its guidelines to create conditions that are acceptable to 80 percent or more of a given space's occupants. The Standard provides a temperature and RH range for summer and winter conditions. Occupants are assumed to be healthy, and the indoor environment is assumed to be "normal." Adjustment factors are provided for different types of clothing as well as different levels of occupant activity, as continuous higher levels of activity increases the metabolic rate and also emits heat into the space.

In addition to finding the most desired temperature and RH levels for the given environment, ASHRAE Standard 55 covers temperature cycling, floor temperatures, vertical temperature difference, temperature drift and other thermal conditions.

We often hear recommendations of 68-78°F as a suitable indoor temperature range for most commercial building types, although there are arguments for narrowing this to 70-75°F. At 78-79°F, some argue, occupants tend to complain that the air "stale."

Saving energy by allowing office space temperatures to rise a mere 3° over 75°, therefore, can lead to IAQ complaints. Others have found space temperatures above even 76°F within the summer comfort envelope to be associated with IAQ complaints in office spaces. Any form of temperature complaint is often mistaken for, or otherwise can lead to, complaints about the overall office environment.

Besides finding the right temperature range, two other concerns include drafts and temperature gradients.

Drafts - Drafts are caused by poor air distribution; solutions include increasing air temperature or lowering air velocity.

Temperature Gradients - Temperature gradients occur when a radiant heat source, such as an electric or steam baseboard heater, produces a higher temperature at the source, whereas a temperature difference can be felt a short distance away from the source. Another example of this is when a warm air supply is carried through ceiling ducts to diffusers in high ceilings; the warm air may cling to the ceiling. Solutions include placement of the return-air grilles at the floor level that pull the warm air down to the occupancy level, or ceiling fans that push the air down.

Sources

Bayer, C.W., Ph.D. and Downing, C.C., PE, "Indoor Conditions in Schools with Insufficient Humidity Control," ASHRAE paper, IAQ.

Beckwith, William R., BS, CIAQ, "Advanced Technology for Economical Humidification to Improve Indoor Air Quality," presented at Why Heat Pipes? Seminar, FL.

Braughman, Anne and Arens, Edward, Ph.D.; "Indoor Humidity And Human

Health," ASHRAE paper, 1996.

Christian, Jeffrey E., "A Search for Moisture Sources," Bugs, Mold & Rot II Conference *Proceedings*, sponsored by the Building Environment and Thermal Envelope Council, 1993.

Dehumidification Handbook, Munters Cargocaire, second edition.

Ellringer, Paul J., PE, CIH, "263 Indoor Air Quality Case Studies in the State of Minnesota," presented at ASHRAE Winter Meeting, 1996.

Flannigan, Brian and Miller, J. David, "Energy-Efficient Dehumidification Technology," Bugs, Mold & Rot II Conference *Proceedings*, sponsored by the Building Environment and Thermal Envelope Council, 1993.

Flannigan, Brian and Miller, J. David, "Humidity And Fungal Contaminants," Bugs, Mold & Rot II Conference *Proceedings*, sponsored by the Building Environment and Thermal Envelope Council, 1993.

"Heat Pipes: A Dehumidification Enhancement," Virginia Power Co. literature.

Ingley, Herbert A., Ph.D., PE, "Case Studies: Applications of Heat Pipe Technology," presented at Why Heat Pipes? Seminar, FL, 1996.

Introduction to Indoor Air Quality: A Reference Manual. Washington, DC: U.S. Environmental Protection Agency, July 1991.

Lstiburek, Joseph; "Humidity Control in The Humid South," Bugs, Mold & Rot II Conference *Proceedings*, sponsored by the Building Environment and Thermal Envelope Council, 1993.

Morey, Philip, Ph.D., CIH, and Swan Jr., Frederick R.; "Microbial Contamination in Buildings: Precautions During Remediation Activities," Clayton Environmental Consultants Inc., Newsletter, October 1993.

Rengarajan, Kannan, "Can Energy Recovery Wheels Mitigate the Impacts of ASHRAE Standard 62-1989 on Florida Offices?", ASHRAE paper, 1995.

Scofield, C. Michael, PE and Des Champs, Nicholas H., PE, Ph.D., "Low Temperature Air with High IAQ for Dry Climates," *ASHRAE Journal*, January 1995.

"Solving Indoor Air Quality Problems in Hot, Humid Climates," course materials prepared by CH2M Hill and University of Florida TREEO Center.

Wheeler, Arthur E., PE, and Olcerst, Robert B., Ph.D., CIH, OSD, Indoor Air Quality, Professional Reference Program, The MGI Management Institute, produced for ASHRAE.

Chapter 7

The Value of
Air Duct Cleaning

Most commercial buildings use air ducts to move heated or cooled air. Most often, we do not see them, do not even know they are there. In this chapter, we will discuss duct cleaning—why it is done and what a facility manager should expect from a service provider. If performed properly as a practical step in routine building maintenance, air duct cleaning can help to ensure a more healthful environment.

PERSPECTIVES

The ductwork in a building is used to deliver conditioned air to the interior spaces. There are many yards, sometimes miles of it snaking through the building's guts. After construction, seldom is any further thought given to the ducts. But in recent years, concern for better IAQ has turned to these internal passageways for air movement. There is still debate over how, when and why to clean inside air ducts. Some claim there still is no need at all, and that air duct cleaners perform unnecessary services to make money.

To be sure, the air duct cleaning industry in its infancy was fraught with fakes and phonies. That is why its reputation has been sullied and many professionals shy away from recommending it or even allowing that is does much good. But that is rapidly changing.

Much of this type of work up until a few years ago was relegated to the residential "blow and go" guys who offered the service for low, low rates. What it amounted to was a quick trip to the basement, some convincing banging on the ductwork and perhaps a once-over with a portable vacuum cleaner. Consumers got, unfortunately, about what they paid for. Just as often, a cheap duct cleaning job was merely a pretense to get into a home and sell other unnecessary products or services, maybe a new furnace, often using high-pressure scare tactics. But in recent years the industry has advanced and gone through its own badly needed internal clean-up. There are always going to be less than reputable "entrepreneurs" entering the business hoping to make a quick buck, as is true with many fields. Today, many trained professional HVAC technicians, the people who had staying power, offer this service, and perform it well, including in regard to commercial buildings, schools, hospitals, apartment buildings and institutions.

Imagine an average-sized office in a building that never gets cleaned. Many duct cleaners use that analogy, comparing a building's ductwork to the size of one average room or office space. Some buildings are decades old, yet have never had their ductwork cleaned. Most duct cleaners have tales to tell about things they found in the ducts, everything from dead rodents, snakes and birds to construction debris and even a tradesman's lunch.

Does cleaning the ductwork in a building resolve IAQ problems? Not necessarily. Duct cleaning is just one more maintenance service that may or may not solve a problem, just as it may or may not prevent one—depending on the nature and extent of the problem. Research shows that it can be a boon. But duct cleaning certainly will not solve any construction deficiencies, such as the absence of a vapor barrier. It will not help if the carpeting is moldy, if the humidity is too high, if drain pans are overflowing, or if chemicals are improperly stored and are off-gassing into a building's air supply.

RESEARCH ON DUCT-CLEANING

Research and studies on the subject of duct cleaning are unfortunately limited. One, reported in the *Annals of Allergy*, involved eight homes in northern Texas sampled to determine

Figure 7-1. A HEPA vacuum designed be 99.97 percent efficient at capturing particles of 0.3 microns in size and up, used in duct cleaning programs. Courtesy: Indoor Environmental Solutions.

fungal colony forming units prior to and after duct cleaning. The authors evaluated the commercial HVAC sanitation procedure of an air duct cleaning company in Fort Worth, TX, to determine its ability to decrease residential fungal populations.

The protocol allowed for sampling prior to the HVAC sanitation procedure to establish a population baseline for each test house and eight weeks post-treatment sampling time to determine the fungal populations' response with time. The study was divided into two phases—one during the winter months when the systems were heating, and one during the summer months when air conditioning was in use. Six houses were tested, admittedly a small sample, with two or more homes serving as a control group. Fungal populations were determined by counting colony-forming units growing on culture media plates exposed to air flows entering and leaving the HVAC system. The houses studied were identified through the offices of three local board-certified allergists, whose patients in this case exhibited chronic symptoms of allergic rhinitis (hay fever) and who had at least one positive skin test reaction to indoor molds. Those who participated did so on a voluntary basis. The HVAC system was sampled by exposing plates directly into the supply air or return air streams, with culture plates exposed for exactly 10 minutes. The cleaning procedure involved first removing the vent registers, taking them outdoors and cleaning them with 0.25 percent solution of glutaraldehyde. A HEPA-filtered vacuum was used to clean out each outlet, along with return air ducts and the HVAC equipment. These were also cleaned with solution, then rinsed and wiped clean. A vinyl-copolymer was fogged into the duct runs using airless spray equipment. Evaporative cooling coils were disassembled, cleaned and reassembled.

A permanent, washable electrostatic air filter cleaning the HVAC system resulted in a significant reduction of mold colony-forming units entering and existing air ducts in this admittedly small number of private residences.

"Heating, ventilation and air conditioning sanitation reduced the number of fungal populations entering and existing the HVAC, suggesting that contaminated HVACs contribute a significant proportion of the total indoor fungal aeroallergen population," the study noted. "Longer-term studies are needed to determine the duration of benefit of HVAC sanitation."

Another study, this one performed by Florida International University, also gave the nod to duct cleaning. Effectiveness of HVAC Sanitation (Duct Cleaning) Processes In Improving Indoor Air Quality was conducted by the Department of Construction Management of FIU through a grant from the state Department of Education, the North American Insulation Manufacturers Association (NAIMA) and the Florida Air Conditioning Contractors Association.

The study sampled eight homes (homes are easier to work with than commercial buildings, but the results should be approximately the same), before and after three different methods of cleaning ducts. All of the homes were of approximately the same age, geographical location, floor plan, building design and materials, and HVAC design and materials. None had ever had their ducts cleaned before.

Climate Control Services, Del Ray Beach, performed the duct cleaning in all cases. "We observed very dirty air handling units, drain pans full of water and debris," according to the study, "and almost clogged filters that have not been replaced on time." Accumulations of dirt and debris were found.

Three methods were tested, including:

- Contact vacuum—Hand vacuuming using commercial-type equipment via existing openings and outlets.

- Air washing or air sweep—A vacuum is connected to the downstream end of the section being cleaned; compressed air is introduced into the duct through a hose with a skipper nozzle propelled along by compressed air.

- Mechanical brushing—A vacuum is connected to the downstream end, with HEPA-equipped negative air equipment used while a rotary brush mechanically or manually dislodges the dirt.

None of the cleaning processes involved the use of chemicals.

All methods were deemed more effective at removing particles of 1.0 microns and larger, which represent allergens such as pollen and mold. There was little effect at 0.3 microns, which in this case were mostly particles traced to tobacco smoke. Unfortunately, the small sample and lack of long-term test results discouraged conclusions on which method actually worked best. Follow-up studies in this area would help.

A lack of information is a major obstacle in solving the problem of sick buildings, the study pointed out. But it suggested that inadequate and ineffective cleaning of HVAC ductwork "may cause and enhance IAQ-related illnesses among the occupants."

See Figure 7-2 for a summary of the Florida International University test results.

Although the study was confined to houses, it pointed out that poor IAQ can be source of allergies and respiratory diseases in occupants. Improper design, installation and maintenance of HVAC systems and ductwork can contribute to this problem. HVAC units can become sources of mold, fungi and other microbial pollutants, Dirt, dust and fibrous material can accumulate inside the ductwork. One way of maintaining the quality of indoor air is to clean the HVAC system and the ductwork.

Still another study was announced around the time this book

HOMES	INDOOR					OUTDOOR			
	PRE	DURING	% CHANGE PRE TO DURING	POST	% CHANGE PRE TO POST	PRE	POST	% CHANGE PRE-IND PRE-OUTD	% CHANGE POST-IND POST-OUTD
CONTROL 1	3095			4641	49.95%	3396	17141	9.73%	–63.07%
CONTROL 2	1114			1865	67.41%	3396	17141	204.85%	–8.10%
CONTACT 1	9369	15796	68.60%	3691	–60.60%	2368	31601	–74.73%	–14.39%
CONTACT 2	3417	19850	480.92%	5076	48.55%	2368	31601	–30.70%	–37.75%
AIRSWEEP 1	5321	15561	192.45%	4715	–11.39%	1714	1904	–67.79%	–59.62%
AIRSWEEP 2	4686	5288	12.85%	2588	–44.77%	1714	1904	–63.42%	–26.43%
MECHBRUSH 1	7290	33373	357.79%	3505	–51.92%	3160	1031	–56.65%	–70.58%
MECHBRUSH 2	6877	8450	22.87%	6296	–8.45%	31601	10311	–54,05%	–83.62%
CONTROL 1	16490			33016	100.22%	21912	40627	32.88%	23.05%
CONTROL 2	563378			10934	–98–06%	21912	40627	–96.11%	271.57%
CONTACT 1	44399	53776	21.12%	27235	–38.66%	19449	44789	–56.19%	64.45%
CONTACT 2	14461	180126	1145.60%	37369	158.41%	19449	447891	34.49%	19.86%
AIRSWEEP 1	28180	208753	640.78%	64917	130.37%	40627	33404	44.17%	–48.54%
AIRSWEEP 2	52028	107044	105.74%	198909	282.31%	40627	33404	–21.91%	–83.21%
MECHBRUSH 1	175612	332913	89.57%	180024	2.51%	49789	207188	–71.65%	15–09%
MECHBRUSH 2	40401	104448	158.53%	231787	473.72%	49789	207188	23.24%	–10.61%

Figure 7-2. Effectiveness of HVAC sanitation at a study at Florida International University. This chart shows the effectiveness of duct-cleaning on particles sized one micron and larger. Readings are average of 15 minutes. Readings were taken two days apart. Courtesy: Florida International University.

was being written, slated to take up to two years, carried out by the U.S. EPA. The first year will study the effectiveness of duct cleaning on a specially designed test facility complete with commercially available air handling unit, supply and return air ductwork, registers, diffusers and controls.

The three common technologies for cleaning air ducts were to be tested—contact vacuuming, air washing, and power brushing. A second year of research was to study nine existing homes, and would include an energy analysis to determine whether or not duct cleaning had any impact on energy conservation or equipment efficiencies.

According to the U.S. EPA, this research was prompted by a growing concern that consumers are sometimes led to believe that cleaning the home heating and air conditioning system can provide many benefits such as improved IAQ, increased energy efficiency, and prolonged system life: "Currently there is little scientific data to support these claims."

Duct cleaning has had its failures, too. Duct cleaning was ordered for the Martin County (Florida) courthouse complex after a number of people complained and some 25 occupants showed symptoms of allergies and hypersensitivity pneumonitis. Occupants complained of itchy eyes, headaches and respiratory problems shortly after two new courthouse buildings opened in 1989. One synopsis of the county's problems read: "March 1992. Ducts cleaned. Other cleaning completed. Occupants' symptoms worsen." It was not for several years, various remedies and almost $10 million later that the building was pronounced "cured."

Figure 7-3. Tools of the trade. Featured is a "duct walker," a brush system that fits snugly into sheet metal, insulated or ductboard ducts and automatically brushes all sides and corners. Courtesy: Indoor Environmental Solutions.

Figure 7-4. Tools of the trade. Featured is a robotic duct inspection and cleaning system with remote control. Inspects, cleans and records the job. Courtesy: Indoor Environmental Solutions.

SAMPLING METHODS

There is debate and some controversy over sampling for airborne microbes. Many facility managers, when they have a problem, expect an IAQ investigator to come in and immediately start sampling to come up with a) the specific cause of the problem, then b) proceed with a solution. Unfortunately, things are not that simple. Many, if not most, reputable IAQ investigators will shy away from in-depth sampling, and with good reason. There are simple, inexpensive sampling programs that will generally verify suspected contaminants and ventilation effectiveness. However, going beyond this to investigate the presence of non-visible mold or fungi organisms can stir debate, as well as the exact methods to sample for such organisms. While sampling for airborne microorganisms is not difficult to do, results are often inconclusive.

Commercial Environmental Systems Inc. (CES), Spokane, WA, uses a method that involves a sampling strip resembling a Band-Aid adhesive bandage, with the pad on one end. This pad contains a growth media for a wide range of organisms. The strip is activated by the inspector, and a one-sq.in. of HVAC duct (or selected building surface) is wiped with the pad. Two strips, each containing a different growth media, are used at each location. These strips are placed in a sterile envelope, which are then returned to the CES laboratory for incubation and analysis.

A report is returned to the client with the results reported in colony counts/sq.in. and on a six-step "severity index" ranging from very low to severe. But experience shows that airborne sampling is not always helpful in identifying the cause of sick building syndrome. For one thing, airborne spores are generally only present in large quantities for short periods of time. Spores being released into the air are dependent on the growing conditions surrounding the organism. If growing conditions are intermittent, as is often the case in HVAC systems, then the release of spores is also intermittent. Only if airborne sampling occurs during the time that spores are being released will significant levels of airborne microbial organisms be found, according to CES.

Surface sampling for microbial contamination, however, can be useful in providing a historical reference of previous growth, and also indicating the potential for future growth. In buildings where high levels of surface microbial contamination have been found, the cleaning and sanitizing of the offending HVAC system or building surface has reduced or eliminated the sick building syndrome personnel complaints. Because of the cost and time requirements, identification of species is *not normally* recommended by CES. Sick building syndrome symptoms are non-specific, and it has generally been found that it is the quantity of organisms present, not the specific organisms, that lead to sick building symptoms. This is different from building-related illness, where species identification may be necessary, and will probably require working with a qualified certified industrial hygienist for complete sampling. If we kill or remove the organisms that are present, it is not necessary to identify them in the first place. Only limited scientific and definitive studies have been done on the relationship between high microbial levels in buildings and occupant reaction, although an increasing emphasis is presently being placed on this research.

With some training and reliable technical support, most knowledgeable HVAC technicians should be able to spot the majority of sick building problems and be able to accurately use relatively simple sampling techniques for verification.

Figure 7-5. Duct-cleaning technicians should wear all appropriate personal protective equipment while on the job. Additionally, all safety precautions should be observed to protect occupants. Courtesy: Ductbusters, Inc.

BEWARE OF MARKETPLACE CLAIMS

Facilities managers should be wary of any duct cleaners who also claim to be IAQ experts. Are they really? Check their credentials. Beware of any exaggerated claims or guarantees for a healthier indoor environment, or sure-fire cures for sick building syndrome. But do not be turned off to an HVAC service that offers duct cleaning as part of its regular routine building maintenance services, or those who claim a good duct cleaning will reduce the amount of airborne dust and particulates in the facility.

Duct cleaning can have other benefits as well, such as increasing the life of filters and the efficiencies of the air conditioning system by improving their ability to move more air versus systems that are dirty and dust-laden.

Demand has helped fuel these advances in air duct cleaning. More manufacturers in recent years have introduced newer, better equipment, just as many duct cleaners have honed their techniques and provided more specialized services. Commercial duct cleaning is no longer considered an unnecessary frill in most quarters, but is a part of sensible building maintenance and preventive medicine. Beware of companies that offer duct cleaning as a service yet have no background in building mechanical systems. Carpet cleaners and others are leaping into the competition, but many of these do not have familiarity with fans, air handlers and other components. Some of these may have to be disassembled during the cleaning process. Some states require a mechanical contracting license to work on (clean) these systems.

A mechanical contractor may be able to spot problems easier, such as a closed fire damper or disconnected, cut or frayed duct liner, and be able to correct them before completing the cleaning.

The connection between dust and allergies is well known. The same study cited earlier in the *Annals of Allergy* describes the effect of HVAC system sanitation on airborne fungus: The presence of climate controls can make the indoor environment more comfortable, and aid in the elimination of outdoor pollens and molds. An HVAC system that is poorly installed or improperly maintained, however, can serve as a primary source for fungal amplification and contribute to the indoor mycoflora.

Based on previous HVAC examination, the authors said they had observed fungal contamination in HVAC systems mainly in the supply air plenum (the area above the unit where the ducts join), evaporative cooling coils, drip pans, and the vent outlets in each room. Little fungal growth is typically found in long duct runs.

NADCA OFFERS HELP

The National Air Duct Cleaners Association (NADCA) was formed years ago to add some sorely needed credibility to the industry. Like other trade associations, its members seek to promote their own business interests, but its members are also better able to communicate with one another to share ideas and marketing plans, as well as disseminate information on the latest duct cleaning practices and techniques. A national certification program and set of standards have come out of NADCA's efforts. This is not intended to be a blanket endorsement of NADCA or its members, or to imply that others who are not members of NADCA cannot ply their trade with equal effectiveness. Simply be aware that standards now exist that you can use to help you in your own awareness of what is required, from a systems point of view, and what is available, from a contractor point of view.

The NADCA certification test consists of 150 questions with a time limit of 3-1/2 hours. As for the nature of the questions, about one-third relate to actual duct cleaning, one-quarter relate to employee and building occupant health and safety issues during the service, and the remainder relate to assessing problems of mechanical systems. To pass, the duct cleaner must have a good working knowledge of HVAC systems, including VAV and CV ventilation, package units, multizones, mixing boxes, dampers, diffusers, sensors, filters, etc. The following are sample questions:

1. When preparing to wet-clean an evaporator coil in place you should FIRST:
    a) replace the HEPA filters in your wet-vac
    b) apply a descaling agent to the coils
    c) make sure that the secondary drain pan has no leaks
    d) ensure that the condensate drain line is working properly

2. How often must a positive/negative fit-check be performed on a half-face negative air respirator?

Figure 7-6. National Air Duct Cleaners Association's Quality Through Knowledge logo. Courtesy: NADCA.

a) each time it is worn
b) quarterly
c) annually
d) semi-annually
e) monthly

3. What percentage and particle size is used when certifying a HEPA filter?
a) 95%, 0.1 micron
b) 97.9%, 0.3 micron
c) 99.97%, 0.3 micron
d) 99.999%, 0.3 micron

The answers are D, A, and C. The certification exam is available to non-NADCA members too, so an air duct cleaner can be a NADCA member but not certified, or vice versa. Again, neither means you are necessarily hiring the best, but it can serve as a useful guideline in selecting this type of subcontractor.

ACCESS POINTS

Any IAQ problem or evaluation should include an evaluation of the ventilation system. Most HVAC systems have few (if any) existing access or entry points for duct cleaning and/or visual inspection. The duct cleaner will have to cut entry holes, then seal them when the work is completed. Typical inspection tools consist of hand-held mirrors, an optical borescope, and possibly a camera or video inspection system. The process then includes following with hand cleaning tools consisting of brushes, either powered or unpowered, scrapers and pneumatic tools such as whips or snakes driven by a compressor. There are various methods used to clean a duct. None is recognized as heads-and-shoulders above the rest; NADCA's own standard recognizes this, dealing only with final results. If practical, ductwork can be cleaned by a technician on hands and knees using a toothbrush. Again, it is the final results that count. But most professionals will use the type of system that does the best job in the least amount of time, which also impacts on the fees they will charge.

Facility managers should be aware that when duct cleaners are on the job, they must protect themselves with appropriate attire. This means, in some cases, full coverage Tyvek-type suits with respirators. When dealing with *Legionella* or asbestos or anything equally dangerous, the service techs will come prepared. Precautions should also be taken to protect occupants, and if alarm among occupants is a consideration this should be done when the facility is empty—which should be done any way, since the ventilation system will have to be shut down, although in some facilities such as hospitals this is not always possible or practical. But any initial walk-through for possible problems, such as a simple inspection with personal interviews, should be done in regular street clothes—we do not need to be spreading any worries or rumors beyond what no doubt are already circulating. This is not the time to be wearing lab coats or jackets with mottoes or slogans like, "We cure sick buildings!" This has the same effect as having a big pink Roach Killer truck parked in the driveway in front of your home. Neighbors start to wonder. Check with the duct cleaner before hiring on this point.

Figure 7-7. Tools of the trade. A 110V vacuum collection system used for duct cleaning. Courtesy: Vac System Industries.

In mentioning hospitals, we should also note that indoor air in health facilities poses special problems, both for the occupants and anyone who investigates or remedies a problem. Precautions must be taken in areas where diseases such as tuberculosis, a highly communicable disease, are found. Transmission can be direct or indirect, and the bacilli are resistant to drying, remaining viable for months. Tuberculosis is an old disease, identified as far back as ancient Egypt and known in Colonial times as consumption. Back then, tuberculosis accounted for as many as one-fifth of all deaths in a year. Despite great strides in keeping it in check, it still kills five million people globally per year. Other, less well known but equally dangerous diseases are also of concern. Source control is important, along with keeping the air flow isolated from other uncontaminated areas, and some type of further air cleaning either from HEPA filtration or germicidal ultraviolet lighting. Cleaning the ductwork can prove difficult, since most are lined and in some cases contain asbestos. Use of robotic equipment may be necessary. At any rate, some firms such as Medical Air Care, part of the Penn Air Group, specializing in IAQ for health facilities and have experience with these types of situations.

VIDEO INSPECTIONS

About video inspection: A number of service companies now stock this type of equipment, and a facility manager may want to specify its use. Rather than simply going ahead and ordering a duct cleaning job that could be expensive and maybe even unnecessary, ask for an inspection first. Using boroscopes or duct-crawling robotic devices mounted with still cameras and/or VCRs, they can give an up-close and personal opinion on the condition of your facility's ductwork. Sometimes problems are simple to resolve: Lack of airflow may be due to something as easily fixed as a closed or stuck damper. Or it may reveal more involved problems, such as a collapsed duct or blockage that would not be helped by a cleaning. This service can also provide ample ammunition for later inquiries on the state of your building's IAQ. Clean ductwork does not mean we do not have any problems, but it means we have at least taken some steps along that road.

Many duct cleaners offer a combination duct cleaning-video inspection service. The video-equipped duct-cleaning tools that are used prove to customers that the job is being done, and show what they are getting for their money. Bob Allen of Clean-Aire Technologies offers the following case history:

The site is a county courthouse in a metropolitan area of the Southwest, 30,000 sq.ft., three stories and housing 40 employees. Air handling units were located in closets and ceilings. A mechani-

Figure 7-8. A video inspection of the ductwork. Courtesy: Clean-Aire Technologies.

cal room is located in a third-floor attic. The ceiling is acoustical tile. The workers have complained about "bad air" for three years. Some employees on the third floor contact local media to complain about suffering from cold-like symptoms and having to work out of their homes. A story breaks in local newspaper and television.

The county purchasing agent contacts companies for a cleanup. When told of color video duct cleaning methods which includes a sample video tape of the air duct cleaning, he hires them on the spot. The systems in this particular case included thirteen 3-10 ton fiberglass lined air handlers with AC coils; 10 reheat coil/ VAV boxes with booster fans; 26 VAV boxes without fans; 114 diffusers; 36-inch lined rectangular main ducts; 4-1/4-inch round, flex and metal feeders; and 3,680 linear feet of ductwork. Specifications include brushing, vacuuming and video scanning in one operation, source removal, application of approved sanitizers to eliminate fungal contamination, and encapsulation as needed.

A crew of three technicians worked over two weeks on the job. One technician was in charge of working the air handlers, fresh air intakes, and supply and return plenums. Two others made access holes and hand vacuumed the main trunk lines and

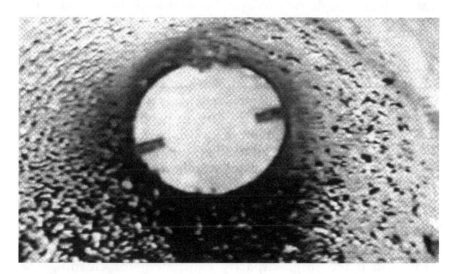

Figure 7-9. The camera's view of a duct coated with microbial growth and dirt. Such contamination is more rare in long duct runs.

VAV boxes. They cleaned the feeder lines. A video tape is made to show that the job was done completely. Cleaning equipment used included the Eliminator 1000 televised duct cleaning system for 6-14 inch feeder lines, one Duct Spy color inspection system, one Eliminator 100 mini-vacuum brush system for flex duct and 4-inch lines, two 110 cfm portable vacuums with HEPA filtration for vacuuming main trunks, Ductmate access doors (for sealing inspection and cleaning holes that had to be made in the ducts), and a low volume-high pressure sprayer for coil and fan cleaning. The coils and air handler were cleaned with a non-acid cleaner. The air handler, coils and ducts were all sanitized with a quaternary disinfectant EPA-labeled for safe use in air systems. The air handler, coils and ducts, after source removal, were sealed with a water soluble vinyl acrylate copolymer using the "Fogmaster" sprayer system.

The original source and amplification site for fungus was located in the air handlers. Some fans and louvers to the fan had a quarter-inch of mold growth. Condensate pans were full of slimy water. Duct liners near the air handling units support mold growth and the sealers used to seal the connections of trunk lines hold clumps of mold similar to that found on the desks. If that was not bad enough, two decaying rodents were discovered in the basement ducts. An unidentified clump of what appeared to be lumpy insulation turned out to be a mound of dead crickets. After vacuuming, their entry points were sealed, along with other open joints and loose fittings. This particular job was billed at a modest 75-cents per square foot, and turned out to be a profitable one for the duct cleaner.

"The good news for the air duct cleaning industry," reports Allen, "is that now we have badly needed credibility and integrity."

Vac System Industries is another supplier of a visual inspection system. Its Super Inspector uses a tracked crawler, high resolution camera and cable to inspect ductwork. A color monitor and VCR can also be hooked up to it, along with various brushes, air brooms and nozzles, a Polaroid camera, boroscope, etc. The company also makes high-capacity equipment for cleaning the ductwork in commercial and industrial facilities. If buying or selecting equipment, the company suggests the following criteria:

- Fan design—determines ability to convey material and create static pressure.

- Fan motor horsepower and voltage—determines ability to continue creating static pressure under the load of dirty filters.

- Pre-filter surface area—affects static pressure loss and the dirt holding capacity of the pre-filter.

- Collection capacity—affects ability to keep working.

- Final filters—affects environment we are working in.

- Filter replacement—affects the cost of operating the portable vacuum collection system.

- Physical size and weight—affects portability.

- Price—determines affordability.

ENCAPSULATION/DISINFECTION

This is not to say that there is not still controversy within the industry. Some duct cleaners routinely add encapsulation and the use of chemical disinfectants to their list of services. Facility managers must be aware of what services they are contracting for, and be sure to stay away from those which are still legally or medically questionable. If the facility manager is only trying to rid the building's ductwork of ordinary accumulated dust and dirt, it is not necessary to specify use of a biocide designed to kill microbiologicals, nor does he need an encapsulant or sealer—any of which could create or lead to even more problems. If microbiologicals are present and application of some sort of disinfecting agent is called for, make certain you or your subcontractor uses only an EPA-approved biocide, one that is approved specifically for use in building air ductwork. Note that if biocides are used that the contractor should take precautions to ensure occupants are not exposed to them directly.

The fiberglass industry at first was cool to the practice of cleaning ductwork. Improper or careless cleaning practices can easily destroy or damage the interior fiberglass lining used in much commercial ductwork. Duct cleaners sometimes responded by saying fiberglass lined ductwork was unsuitable and prone to such damage. Working together, the two factions have resolved some of their differences. As a result, the North American Insulation Manufacturers Association (NAIMA), a trade group of fiberglass suppliers, came out with its own standard and recommendations for cleaning fiberglass lined ductwork. NAIMA also has endorsed workplace rules for IAQ proposed by OSHA, which in part calls for products such as fiberglass to be tested for safety during the development and manufacturing stages. NAIMA said it was confident that its product testing could "refute the notion that these products may degrade indoor air quality. In fact, industrial hygiene studies measuring airborne fiber levels in buildings found fiber levels at or below those present in the outdoor ambient air." Individual fiberglass manufacturers have also come up with several new products which have a surface coating that is more durable and more easily cleaned than liners of old. And some companies say they have had success in encapsulating damaged coatings—Vac Systems International says its "Tough-Coat" product enhances and repairs mechanical insulation, applied after source removal of dirt and other debris. It also contains an antimicrobial agent.

The building in this case was a 15-year-old, five-story office building which won an architectural award for energy conservation. It had two types of ventilation system for 46 air handling units: mixing boxes joining cold and warm air decks in some places, no mixing boxes in others. Both round and rectangular ductwork was used. The round duct was externally lined, while much of the rectangular duct had an inner liner. Filtration was judged adequate, with 25-35 percent pleated pre-filters and final bag filters. Nevertheless, occupants began complaining of headaches, respiratory problems, etc. It was discovered that the miles of ductwork had extensive microbial contamination. A number of short-term solutions were attempted, including cleaning the ductwork, but the contamination returned. Coatings were tried with little success, and attempts to reduce the moisture in the ductwork also met with failure. The final solution was a combination cleaning, application of an EPA-registered biocide, and use of the mechanical insulation repair coating to the internal lining. Follow-up checks in the form of wipe samples for colony-forming units showed dramatic results. In one stretch of supply side duct, 65,000 colony-forming units initially gave way to only one colony-forming unit. Another revealed a 240,000 colony-forming unit count that was reduced to 3 colony-forming units. It should be noted that not everyone agrees on the use of coatings in ductwork. Some coatings may not do the job, and some may negate the fire resistance ratings of the liners. Caution should be used in this area, and use of a coating should be considered only after initial cleanings and other remediation efforts have failed.

ODOR INJECTION

Switching tracks slightly, some people are hard at work putting things into ductwork that you may not have been aware of, or even suspected: scent. In the same sense that scratch-n-sniff

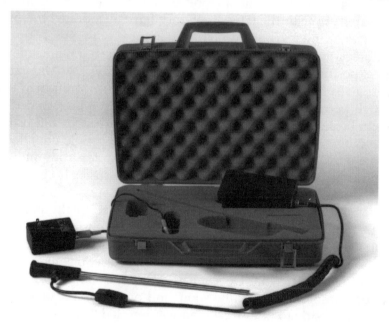

Figure 7-10. A boroscope used for "before and after" records—ductwork inspection and to demonstrate the work has been done satisfactorily. Courtesy: Indoor Environmental Solutions.

advertising gimmicks and scent strips in magazines are used to sell items like perfume, comes scented air ducts. Two Marriott hotels in Florida have reportedly experimented with introducing fragrance to the ventilation systems in their lobbies. One has done it for two years, reporting no problems, while the other said it scrapped the idea due to high cost and some complaints. Retailers, likewise, are experimenting with scent to produce a friendlier, fresher-appearing area that will induce people to stay and buy products. Some of these scents are conveyed by the air duct system, and one consultant said he foresees the day when adding scent to a room will be as commonplace as air conditioning or heating it. Inducing a fresh bakery smell to a shopping mall food court is one example of the retailing side, but it can also extend to new car showrooms (taking advantage of that much-vaunted "new car smell") to masking unpleasant formaldehyde odors in furniture stores, or animal odors in pet shops.

Some groups protest this unwanted invasion of fragrance into our society. Department stores once bent on spritzing incoming customers with the latest perfumes have abandoned the practice after one large department store was sued by an unappreciative customer. Some employers make it a work rule: no strong perfumes, colognes, aromatic hair spray, etc. Magazines occasionally hear from unhappy readers who must sniff their way through perfumed ads. Some magazines now require scented inserts to be sealed, to leave it to the reader whether to sniff or not, while others will not accept them at all. Some people are allergic to such odor additives, while others claim to suffer from multiple chemical sensitivity brought on by exposure to such chemicals. The National Medical Advisory Service held a state-of-the-science symposium to help sort out whether or not there really is such a thing as multiple chemical sensitivity. The symposium addressed this issue with such talks as "Indoor Allergens and the Diseases They Cause" and "Human Hypersensitivity Reactions to Environmental Chemicals: What Do We Know?" It is a complex issue not due for a resolution any time soon, and one with a legal as well as a medical side to it. One session was entitled "MCS: The Influence of the Courts on Public Policy."

Scented ductwork should present no problems as long as the facility manager is sensitive to reaction from occupants. But do not use such fragrances or scents to mask another problem, especially one that could be hidden or potentially serious: such as the growth of mold and mildew. Better to, as duct cleaners say, attack the source rather than to mask it with a diversion.

DUCT CLEANING CHECKLIST

The U.S. EPA offers 10 tips on duct cleaning:

1. Duct cleaning should be scheduled to take place when the building is unoccupied to avoid exposing occupants to chemicals or particles shaken loose during the cleaning process.

2. Use a vacuum wherever possible and keep it on to avoid drift of particles into occupied areas.

3. Take measures to protect the ductwork if the ducts will be cleaned. If new openings are to be created, follow the manufacturer's directions for sealing. Large vacuums should be used with care, as high levels of negative pressure combined with limited airflow can collapse ducts.

4. If using high-velocity airflow to clean the ducts, be sure to include controlled, gentle brushing of duct surfaces to dislodge particles.

5. Use only HEPA vacuums in occupied spaces.

6. Use only EPA-registered biocides when biocides will be used to kill microbes, and follow the manufacturer's directions carefully.

7. It is generally not recommended to use sealants to cover interior ductwork surfaces.

8. Clean and sanitize coils and drip pans to reduce microbials.

9. Remove and replace contaminated porous materials (such as fiberglass duct liners) in the ducts and air handler.

10. Implement a scrupulous planned maintenance program after cleaning.

Sources

Bas, Ed, *Indoor Air Quality in The Building Environment*, Business News Publishing Co., 1994.

Groen, Doug, "Mechanical Insulation Repair Coatings: A Case Study," Healthy Buildings, National Coalition on Indoor Air Quality (NCIAQ).

Krell, Bob, "HVAC System Hygiene for Health Care Facilities," Healthy Buildings, National Coalition on Indoor Air Quality (NCIAQ).

Montz, W. Edward Jr. Ph.D., "Indoor Air Quality in Health Care Settings," *Plant, Technology & Safety Management Series*, No. 3, Joint Commission on Accreditation of Healthcare Organizations, 1995.

Scott, Richard, AIA, "The High Cost of IAQ," *Engineered Systems*, Business News Publishing Co., January 1996.

Yacobellis, Tom, "Introduction to HVAC System Cleaning Services," Healthy Buildings, National Coalition on Indoor Air Quality (NCIAQ).

York, Aaron, "Duct Cleaning: A Checklist of What, And What Not to Do," *Contracting Business*, October 1995.

Chapter 8

Putting It All Together: The IAQ Program

This is where we put all of the knowledge we have learned in this book together and begin an IAQ program. Whether there is a perceived problem with IAQ or not, this is the time to start, as prevention is as important as a cure. IAQ practices need not be very complex, nor very expensive. Most problem solutions do, however, require a big picture view of the mechanical system, a thorough clean-up and ongoing maintenance.

IAQ MYTHS TO AVOID

A commonly held myth of proper IAQ is that if a building meets ASHRAE Standard 62 for ventilation, then it will not have indoor air problems.

Another myth is that by maintaining proper temperatures, there will not be problems with humidity. The reality is a bit more complex than this.

Actually, a broader, four-pronged approach is needed to achieve the proper IAQ mix for a building:

1. Identification, location and elimination of the pollution source, and/or control of the pollutant pathway (which may be the HVAC system).

2. Adequate ventilation, including proper quantity ventilation rate, air balancing/pressurization and air distribution.

3.   Adequate filtration.

4.   Adequate moisture and humidity control.

Leaving out any one of these key ingredients creates an imbalance which is hard to remedy merely by increasing or enhancing the other three. When one is out of sync, it is hard to orchestrate the remainder of the chorus.

For example, moisture loads from outside air can often exceed the dehumidification capacity of the building's HVAC equipment, just as during the summer, cooling (and especially moisture removal) requirements can exceed the HVAC system's capabilities.

Source removal, often considered the best way to rid a building of pollutants, sometimes is not possible. Bringing in more outside air to dilute these pollutants is another good idea, but can be expensive because we have to cool this air as well as dehu-

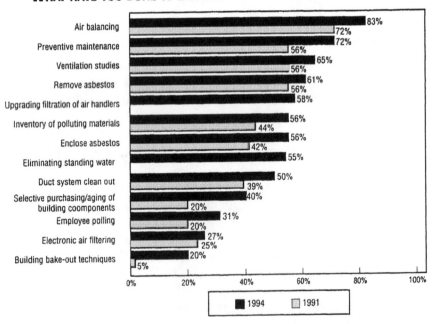

Figure 8-1. IFMA survey on IAQ. Courtesy: IFMA.

midify it. Filtration also helps, but not if the filter system is going to be easily overwhelmed by the load of pollutants we are trying to exclude.

Instead, using a combination of these efforts is necessary. Provide proper interior pressurization, exclude excess outside moisture through envelope modifications, and use dehumidification equipment where needed. These efforts can pay off in better air quality, higher productivity, as well as longer life for the building's materials and furnishings.

Focus On Microbial Contamination

"Dilution is the solution to pollution" is a common theme of many IAQ programs, leading to greater introduction of air to dilute indoor contaminants. But in the case of microbial contamination, we can often get at the source and solve the problem there. Some researchers say that the extent of IAQ problems related to microbial contamination has previously been quite underestimated. Some attribute this to a "chemical materials bias" and to a lack of interaction and understanding between the environmental health and microbiology communities. But microbes are getting a lot more attention, particularly by OSHA, which has proposed regulations on indoor air to address microbial contamination via moisture control and contaminant removal.

Enviros: The Healthy Building Newsletter, calls this a "paradigm shift towards an interdisciplinary, building wellness approach to public health and building problems."

A number of indoor air investigations and case studies help to demonstrate the impact of microorganisms upon the quality of the indoor environment:

- Recent studies describe microorganisms as the primary source of symptoms in as many as 35-50 percent of IAQ cases.

- The Minnesota Department of Employee Relations reports that approximately 33 percent of IAQ problems are related to microbes.

- In the U.S. EPA's *Building Air Quality Guide,* five of the 15 (33 percent) "Sample Problems And Solutions" are microbial examples.

- In OSHA's proposed IAQ standard, 24 of the 73 pages (33 percent) from the preamble and standard contain references to microorganisms.

HIRING AN IAQ CONSULTANT

Residential duct cleaning may be a fairly low-tech, low cost job that has to be done right but does not require enormous skills or effort. Many companies are getting into this business because of that: The initial capital investment outlay isn't large, employees can be found and trained in a few days or at least weeks, and wages paid out in this business are usually relatively low. A thorough commercial job, however, can be much more complex and demanding. Tom Yacobellis reports that his DuctBusters Inc., Clearwater, FL, company was awarded a contract for $250,000 to clean the ductwork in the Polk County Courthouse, a sick building that had undergone some $30 million in remediation efforts from 1992-94. This was going to be one of the final steps in a long chain of getting this building open and putting an end to employee complaints that stemmed from the time it opened in 1987. However, in an about-face, the county board decided to forgo cleaning most of the ducts, opting instead to replace much of the ductwork instead at an additional cost of $3 million. This was not necessarily an indictment against the practices or methods of cleaning commercial ductwork. Rather, it was an attempt to sidestep any possible future complaints that could cost the county millions of dollars in the form of lawsuits filed by those who claimed to be afflicted over the years.

One highly successful duct cleaner predicts the number of competitors in his Midwestern market to double within the next few years. He sees the demand also continuing to grow. His business is solely duct cleaning, not HVAC. He claims this is an advantage in that suspicious purchasing agents do not see his business as merely a ploy to sell HVAC equipment or services. Nor do HVAC contractors see him as a threat—if they do not currently offer this service, they are usually more than happy to refer their clients to him. This same duct cleaner said what does alarm him is the number of new competitors cropping up who have no backgrounds in HVAC—maid services, chimney sweeps, etc. are all looking at this as a possible growth industry. The facility manager's duct cleaning company should have a working knowledge of his building's HVAC equipment. The two really go hand in hand. HVAC equipment must be cleaned at the time the ducts are cleaned to ensure a complete job. Otherwise, the dirt remaining in the equipment will merely be redistributed throughout the system when it is turned on again. If nothing else, a company that is hired for duct cleaning must be able to go in and clean the ducts without damaging the coils and other components of the HVAC system they will be working on. Find out how much experience the firm has in commercial work—to many, this is a quite new enterprise, and they may not have either the equipment, expertise or experience to handle a sizable commercial job.

One large mechanical contracting company sees IAQ as potentially as large an issue as elimination of CFC refrigerants. Lawsuits and fear of lawsuits, rather than OSHA, are driving companies to take a closer look at their indoor environments. The company has been specializing in air flow and air movement stemming from work it has done for clients such as pharmaceutical laboratories. Another is focusing on IAQ as an adjunct to its background in air balancing. Again, the trend seems to be a flow of information from specialized mechanical systems such as hospitals and laboratories to the more mainstream, such as schools and office buildings. The latter company takes an approach to IAQ that includes basic mechanical air balancing, fans, motor RPM, ventilation including cfm requirements under ASHRAE 62, percentage of outside air, damper operation, cooling water temperature and pressure, etc., providing the necessary documentation to meet OSHA requirements. Most companies will outsource some of this work, such as duct cleaning if it is needed. A typical service package might include all phases of mechanical HVAC projects from design analysis to start-up and commissioning.

Some experts estimate that an initial IAQ investigation can run $3,000-$7,000. The odds of finding a bargain for less are not great, but a higher price does not guarantee a good job either.

Travis West, an IAQ investigator in The Woodlands, TX, addressed the BOMA membership at one annual convention in Denver and published a paper titled, "How To Select And Evaluate Indoor Air Quality Consultants." Some tips from his presentation are offered below. Above all, strive to be an educated consumer.

1.   Ask first what the work will involve. How will the consultant handle inspection, ventilation system inspection, contaminant measurements, data interpretation and mitigation?

2.   Ask for a sample report from another job, so as to see what areas will be covered. Do not expect a prospective IAQ investigator to be able to provide names of past clients as references. Most clients will want to remain confidential. If necessary to retain anonymity, the name of the client on these sample reports can be blocked out. Ask prospective consultants' questions such as what they will attempt to measure, how they will interpret the data and how they have mitigated similar problems in the past.

3.   Interview building occupants as this will assist in developing a test protocol if there is already an IAQ problem. The follow-ups should be conducted in person, not just by questionnaire, as other issues will crop up that will help narrow the investigation.

4.   Be cautious when using an HVAC contractor as an IAQ consultant. An HVAC contractor, while ultimately useful in any IAQ investigation, may be more likely to recommend investing in new or replacement mechanical equipment as a means to solve the problem.

5.   Be wary of any IAQ investigator who claims to have been in the business for 25 years—it is still a relatively new field. By the same token, the facility manager should not necessarily be turned off if somebody says his company has been at it for five years or less.

6.   Look for a company that has regional expertise. If the facility is located in the South, somebody with a background only in the Northwest is not going to do much good, since IAQ problems are often specific to a region or climate.

7.   Beware of ventilation system consultants, duct cleaners, industrial hygienists and environmental firms with narrow fields of expertise, such as asbestos abatement, hazardous waste and underground storage tanks. They might not have the experience to handle commercial building problems. Note: Newcomers to this rapidly growing field will be especially open to professional malpractice lawsuits.

8.   Consider using an attorney to hire an IAQ consultant. Dealings between the two may be considered confidential, customer to client, and may be kept out of a court of law (no guarantees)! Do not use the building owner's present attorney, who might be viewed as an employee.

MCS Is Compensable, Says Court

One employee of a gambling casino in Reno, NV reportedly developed multiple chemical sensitivity after a quantity of pesticides were sprayed near her work area, including into the ventilating ducts, in an attempt to control cockroaches.

Up to 100 people ultimately were affected by the spraying enough to require medical attention, but this one woman was affected more than most. She made repeated trips to hospitals to control symptoms including dehydration brought on by uncontrolled vomiting. She and other casino employees successfully sued for workers compensation benefits. The Nevada Supreme Court ruled that multiple chemical sensitivity—although the existence of multiple chemical sensitivity is still much in debate in the medical community—is indeed compensable. The court ruled that multiple chemical sensitivity was in this case demonstrably tied to the plaintiffs' chemical exposure at work.

THE IAQ PROGRAM

This section offers the reader tips and steps for implementing an IAQ program to be used, if desired, in addition to the information provided in the other chapters that specifically address various aspects of HVAC performance. This section was compiled from a number of sources: the U.S. EPA; the Business Council on Indoor Air; Jay Kirihara, VP of QIC, a Dallas-based IAQ mitigation firm; the American Industrial Hygiene Association; BOMA International; Jack Halliwell, an IAQ consultant and professional engineer; and Michael A. Price, an IAQ consultant.

Be Proactive

The road to healthy buildings should not start when someone lodges a complaint, or in response to a threatened lawsuit. The time to start is now, and the easiest way to get there is by doing what is being done now by reading this book—becoming educated and informed on the issues. Facility managers who allow no surprises, let no scare tactics or uncommon terms confuse them, are already on the right road. "An apple a day keeps the doctor away" takes some of the myth out of staying healthy, does it not? Same with: Exercise, eat well, drink plenty of fluids and get enough rest. Maintaining a healthy building need not be much more mysterious. Avoid shortcuts, develop a routine maintenance practice and stay with it, be aware of some of the dangers that could be lurking around and they will be that much easier to avoid.

"Don't wait for the sirens," advises Richard R. Monsen, PE, president and CEO of Monsen Engineering Co. "Taking a proactive stance worked for us with CFC management. We didn't wait for the government to give us a timetable then, and we do not advise our clients to do so now. Preventing 'sick building' problems makes much more sense than fixing them later. And it can be done." Monsen's approach is to conduct a thorough IAQ audit to uncover potential problem areas and to prove healthy areas are *really* healthy. Extensive documentation by a responsible firm not only demonstrates to employees that the employer takes the issue seriously, it also provides powerful evidence against future legal problems.

W. Edward Montz Jr., Ph.D. of Indoor Air Solutions Inc. of Pottstown, PA, calls being proactive on IAQ a "good business decision." He added, "We have calculated that clients who have IAQ problems expend 2-10 times more money solving problems than they would have expended on a proactive program."

The Building Profile

The first step in an IAQ problem is to develop a building profile. It should include information on:

- Construction and operating documents.
- Commissioning.
- Test and balance.
- Remodeling or renovations.
- Equipment descriptions.

The Investigation

The IAQ investigation should answer:

- What are the temperatures in different spaces, on different floors, in various parts of the building?

- What is the humidity level in those spaces? If the humidity levels are high, check for mold or signs of moisture on carpeting, furniture, walls and ceiling tiles.

- What are the CO_2 concentrations in those spaces. If it is over 1,000 ppm, there may be a ventilation problem.

- Are there any tell-tale odors?

- Is tobacco smoke a problem? Are there complaints? If there is a designated smoking area, is it exhausted separately from the rest of the floor or space?

- Are there complaints about acoustics, lighting or other environmental conditions? These stressors could be confused with poor IAQ. Check the design for sensory stress (olfactory, auditory, visual and thermal) and for psychological stress-

One Sick School

One case history of a school in Hagerstown, MD, illustrates well how shrewd investigation and follow-up to complaints can solve a problem, but also how the answer is not always neatly or conveniently arrived at ("Cure of a Sick Building: A Case Study" by Knud Helsing, Charles Billings, Jose Conde and Ralph Griffin, Pergamon Press, "Environment International"):

Northern Middle School was built in 1979, a two-story brick-and-concrete structure designed to be energy-efficient. Soon after the building opened the staff began complaining about poor air circulation, odors, headaches, thermal discomfort, etc. An engineering consultant recommended several changes. Among other things, he found that the stack for an oil-fired furnace was not high enough, allowing some exhaust gas to reenter the building. But the problems continued. Actually, when student and staff absenteeism records for the past five years were studied, it showed no increase from the time the students moved into this new school from their old one. Nor did academic records seem to be affected. Still, five out of six faculty members reported headaches and fatigue, most often late in the day. CO and CO_2 levels were monitored, cleaning materials were surveyed and a smoke test of the sewer system was done. The latter revealed several leaks, which were immediately sealed. Several changes were also ordered for the ventilation system, even though CO and CO_2 levels were not considered a problem. Larger VAV units were installed in several areas, and fan speeds increased to move more ventilation air. Supply air temperature was lowered, from 58° to 50°F, in what was originally an energy-saving effort. Pockets of stagnant air were eliminated, and bathroom exhausts were installed.

These changes led to an almost immediate, dramatic fall-off in complaints about the building. Even more important, specific complaints related to headache, asthma, earache, dizzy spells, fainting, nosebleeds, etc.—all of which were carefully documented over a two-year period—declined appreciably. The changes made in this case were relatively simple and easy to make, but they were also synergistic. Following only one or two of them might or might not have yielded worthwhile results. In addition, the school board continued to investigate even though some factors, such as CO and CO_2 levels, as well as absenteeism and academic records, seemed to indicate there was no serious problem.

Lessons: Take all complaints seriously, and do not overlook obvious sources of contaminants, such as the sewer line leaks. Nor can we assume that even a new building has proper HVAC equipment and acceptable air mixing.

causing conditions. (Would you want to work/live in that place?)

- How was the building originally intended to function? Consider the building components and furnishings, mechanical equipment (HVAC and non-HVAC) and occupant population and associated activities.

- Is the building functioning as designed? Find out whether it was commissioned. Compare the information from the commissioning to its current condition.

- What changes in building layout and use have occurred since the original design and construction? Find out if the HVAC system has been reset and retested to reflect current usage.

- Has the use of the space changed so that the existing mechanical equipment cannot handle the need? Increasing the number of occupants in the space, longer work hours and changing how the space is used can be too much for the HVAC system to handle. Office partitions may be disrupting airflow.

- Have there been any IAQ problems in the past? Consult records to determine if there is a history of IAQ problems.

- Are there current occupant complaints? Review records and interview the occupants.

- Are occupants using personal humidifiers? This may be a sign of a humidity problem. Are they being maintained?

- Do occupants routinely get shocks due to static electricity build-up? This means the air in the space is very dry.

- Are occupants attempting to change air distribution by taping cardboard over supply air grilles? This indicates dissatisfaction with the air distribution or temperature.

- Is it warm near the ceiling but cold at the occupant level during the winter? This means warm air from the supply grille is sticking to the ceiling level and not reaching the occupied level. A return air grille may need to be installed at the floor level or other measures taken to draw the air down.

- Inspect the ductwork and HVAC system, particularly the air handler, for microbial contamination, moisture, excessive dirt and malfunctioning or inoperative components.

- Develop a pressurization profile of the air handler to determine both the effectiveness and efficiency of the system.

- Inspect the inside and outside of the building for possible sources of pollution.

IAQ consultant Michael A. Price suggests that every building IAQ investigation begin with the following to see if any of these conditions are present:

- Outside air intake adjacent to loading dock.
- Outside air intake near heavy motor vehicle traffic.
- Nearby construction activities contaminating outside air.
- Exhaust or industrial stacks near outside air intake.
- Outside air intake adjacent to exhaust or relief air.
- Outside air intake adjacent to cooling tower.
- Roof ponding or standing water near outside air intake.

Even Hospitals Get Sick

One might think that hospitals and similar medical facilities are among the cleanest places to be, but hospitals suffer from some of the same illnesses as other facilities, including deferred maintenance, mechanical systems that fail to operate according to plan, a failure to recognize problems early on, etc. Much of this is brought on by increased attention to the bottom line, as hospitals feel the crunch from public watchdog agencies and insurance companies. The addition of hi-tech electronic diagnostic equipment and computers has added to the heat load, as well as the heat from more and more medical specialists showing up in the operating rooms. At the same time, medical personnel are covering up with more layers of protective clothing to avoid contact with tainted blood from communicable diseases such as AIDS. In these facilities, errors and omissions in the HVAC system can add up to disaster for patients and employees alike. Containing highly contagious diseases and protecting patients from secondary infections during their hospital stays is contingent on having mechanical systems that operate according to design, from start-up through what is typically many years of operation. The following case history was submitted as the subject of a technical paper prepared for an "IAQ and the Law" workshop, ASHRAE IAQ '95. The project was a 10-story addition to an existing hospital.

The HVAC system was a dual-duct system with 17 air handlers including an outside air pre-treatment unit; five return air fans; 26 exhaust fans; and 505 terminal units and pneumatic controls.

The owner elected to expand the facility in 1977. Construction began in 1981, with substantial completion of the project in 1983. Occupancy of the addition took place in 1984-85. In 1986, the owner noted problems maintaining the required positive air pressure relationship between the surgical suites and exterior corridors—crucial in order to isolate highly sensitized patients from non-sterile daily routines and traffic. An air balance contractor was hired, but was unable to balance the HVAC system because of air quantity discrepancies, resulting in a retrofit of the terminal unit control scheme. Problems continued, with a "medical emergency" declared in 1987 brought about by this continued inability to isolate the surgical

Even Hospitals Get Sick (*continued*)

rooms, followed by another new control scheme and 74 new terminal units installed on the surgical floors. Nevertheless, problems continued, with another expensive ($1.8 million) mechanical equipment retrofit ordered in 1988, to a single-duct terminal reheat system. The owner sued the architect, the design engineer and the general contractor in 1989, asking $1.8 million in damages plus $2.4 million in lost revenue and other expenses. The lawsuit was settled two years later for $600,000.

Richard J. Tyler, partner in Jones, Walker, Waeschter, Poitevent, Carrere & Denegre, LLP, New Orleans, who prepared the case history, said an examination of the system showed that the original system was designed to be marginal, incapable of delivering the specified number of air changes and maintaining the static pressures necessary to operate the terminal units (it was, however, capable of delivering the required static pressures in surgical suites, labor and delivery suites). The inspection also showed:

Figure 8-2. Reprinted with permission from *The Heating, Refrigeration and Air Conditioning News*, Business News Publishing Co., Courtesy: Dan Saad.

Even Hospitals Get Sick (*continued*)

- A broken damper linkage resulting in the outside air damper to an air handler being 90 percent closed.

- Improperly set smoke purge damper linkages which directed large amounts of return air to the outside rather than to the air handler.

- A blocked bird screen for the outside air pretreatment unit.

- Air handlers running at less than full-load amperage.

- Dirty air filters for the air handlers (lint from massive laundry operations are always a problem, often clogging filters and positively pressurizing the laundry facility, thus pushing contaminated air out into adjoining areas).

- Missing fluid in an air handler manometer (used to indicate when replacement is needed).

- Few access doors for ductwork inspection.

- Broken turning vanes in the ductwork.

- Internal duct insulation that was too loose.

Other items were identified, including a lack of training experience for hospital maintenance personnel and a lack of written materials to aid them, such as equipment manuals for maintaining the system. But the point of this story, Tyler noted, was really in the lawsuit that never happened: The possible litigation that could have been filed by patients and/or hospital employees affected while all of these "improvements" and investigation was going on. Often, a "cure" has to be first implemented and then studied for a time to confirm its effectiveness, or lack thereof. Other solutions must be sought if one fails. Certainly, this was no isolated case, either. None of the problems was unusual, but many of them together contributed to a synergistic effect that was not only more difficult to cure, but equally difficult to diagnose. More cost-effective solutions must be tried before full-scale equipment changeouts can be ordered. Often, buildings or at least some of the conditioned spaces in them must continue in service while the "cure" is studied.

> **Even Hospitals Get Sick (*continued*)**
>
> This is sometimes out of economics, but also because few results can be accurate gauged if the conditioned space is unoccupied. Tyler points out that "IAQ lawsuits related to poor operations and maintenance of HVAC systems will not stop at the building owner; design professionals who know or should know of such problems and fail to address them in system design will be implicated as well."

The building's air filtration system must be examined for fit, efficiency, current condition and replacement schedule. In addition, the following items should be checked:

- Dirty humidifier reservoirs or cooling towers.
- Poorly draining condensate pans/trays.
- Drain pans pitched improperly.
- Lack of or improperly designed traps.
- Torn and shredded insulation.
- Damp internal insulation.
- Rusting of internal surfaces.
- Mold contamination on internal surfaces.
- Improperly maintained dampers, actuators or linkages.
- Broken mechanical linkages on dampers or controls.
- Poorly cleaned and maintained pneumatic control systems.
- Fans and blowers wired incorrectly.
- Loose or broken fan belts.
- Materials stored within air handling equipment.
- Inoperable motors.

Action Steps

There are ways and means of keeping our buildings healthy. Some are logical extensions or merely adherence to regular main-

Figure 8-3. Renovation and expansion activities can expose occupants to products or particulates of construction.

tenance practices. Others are an awareness and attention to specific woes, such as sources of pollution or too-high humidity. Still others can be enhanced by selection or use of proper HVAC equipment.

IAQ problems can be difficult to diagnose. Preventive medicine can go a long way in avoiding problems before they occur. Jack Halliwell, who has helped resolve hundreds of IAQ problems in office buildings, says: "The first principle is that IAQ problems, by nature, are multifactorial. In other words, IAQ problems are created by a *combination* of conditions (the interaction of a number of building related problems) occurring *simultaneously*. Finding the cause of an IAQ problem can be similar to solving simultaneous equations, and in this instance, each equation represents a different building system. HVAC, automatic temperature controls, operations and maintenance, renovations and the building envelope itself can all contribute to, or even create the problem."

Furthermore, IAQ problems seldom affect everyone in a building, or affect everyone in the same way. This leads to skepticism over whether there is indeed a problem at all. And problems caused by other factors, such as noise, lighting or job stress, can be even more difficult to pinpoint. There is no Swiss army knife with all the answers to IAQ problems contained in one handy gadget.

Third, Halliwell points out, is that "a significant amount of

Figure 8-4. Part of the investigation process should include inspection of the exhaust vents and air intakes to ensure they are not located too close together, resulting in the building feeding bad air back into itself.

similarity exists among the health symptoms caused by *different* pollutants. With different pollutants causing similar reactions in people, it becomes enigmatic to diagnose the cause of the problem based solely upon an occupant's physical symptoms."

As always, prevention is best. The expense and effort required to prevent most IAQ problems is usually much less than the expense and effort required to resolve problems after they develop.

Here are a few action steps, both general and specific, that can be taken to prevent or resolve an IAQ problem:

• Become familiar with HVAC operation and maintenance, particularly the makeup of the building's current HVAC system, and IAQ.

Middle School Cleans Up Its Air

A middle school in Washington State was closed for a period in 1994 when there were complaints and illnesses reported by students and faculty. Headaches and upper respiratory irritation had been reported at the school for an extended period of time. In April 1994, there were complaints of nausea, headache and dizziness—in other words, the very symptoms that often are attributed to some fault or flaw in the indoor air or environmental quality.

The school was ordered closed, and while no unusual contaminants were immediately discovered, it was decided to act—some old carpeting was removed, and the HVAC system of the school was given a thorough cleaning by a mechanical contractor. It was also decided to monitor CO, CO_2, temperature and relative humidity, VOCs, formaldehyde, particles and biocontaminants in the search for a solution.

Air Quality Sciences Inc., Atlanta, GA, performed the IAQ investigation. As a result of the complaints, the school district also moved to establish an Indoor Air Quality Committee consisting of school faculty, administration, staff, parents and community representatives.

Internal fiberglass insulation lining the HVAC system ductwork was removed. After carpeting was replaced, new carpet cleaning and maintenance procedures were established.

Communication was considered a key factor in the remediation process. The IAQ committee established procedures to relay IAQ information to the concerned parties. Those procedures included: providing letters from the principal to all parents reporting the status and plans of all building and health monitoring; keeping the local media informed via building and health reports and public meetings; keeping logs of commonly asked questions and providing responses to those questions through the monthly school/PTA newsletter; and providing reports to the school board, PTA council, employee associations and other groups, as needed. A response form was provided to all school staff to track concerns, responses and solutions. Unusual odors or uncomfortable building conditions became "reporting mechanisms" for IAQ concerns.

Again, it was difficult to pin down one single cause for the

Middle School Cleans Up Its Air (*continued*)

complaints in this case. Comparison of absence rates at the problem school and a control school showed that the rates were not that much different, and some complaints such as headache and stomachache were similar at both problem and control schools.

After the school was reoccupied, follow-up testing "indicated that a normal, indoor environment existed."

- Make IAQ a priority: Institute health, open communication, and respect as priority principles for building operators and occupants.

- Practice good IAQ management: Prevent pollution by implementing good IAQ management practices.

- Fix things that go wrong: Establish effective problem-resolution procedures and act to solve problems promptly.

- Appoint an IAQ coordinator and train him and other personnel, such as the maintenance staff and the human resources manager, on IAQ and IAQ procedures. The IAQ coordinator does not necessarily have to be a person involved in building maintenance, but must be someone who is familiar with procedures including operation of the building's HVAC system. Usually, this will be the facility manager.

- Provide occupants designated smoking areas or abolish them entirely. If a smoking area is designated, it must be properly ventilated with a higher rate of airflow to dilute smoke particles. It should also be exhausted separately from the rest of the floor or space.

- Follow up on problems to ensure they are solved.

- Do not discount complaints as being "just in the head." Take each IAQ complaint seriously, and document all complaints.

- Consider what changes may be needed to prevent IAQ problems from developing in the future. Consider potential changes in future uses of the building.

- Institute a good cleaning regimen or maintenance program for the HVAC program.

- Conduct a feasibility study of incorporating retrofit or replacement of existing HVAC components with energy-efficient components to reduce energy costs.

- Store chemicals properly and use pesticides judiciously. Cleaning compounds and chemicals should be stored tightly capped, preferably away from occupied areas. The same goes for gasoline, solvents, paints, insecticides, etc. Avoid insecticide use near outdoor air intakes.

- Air new carpet and furnishing prior to installation or before exposing occupants to them.

- Observe good housekeeping practices such as floor or carpet cleaning.

- Improve and properly maintain the building's ventilation, filtration and moisture and humidity control systems (see Chapters 4, 5 and 6).

- Consider employing an IAQ consultant for investigation and mitigation services.

- Avoid potentially offensive building and maintenance materials. Check with manufacturers for VOCs and formaldehyde content.

- Design, construct, and operate with adequate ventilation.

- Design and operate cooling and heating systems conservatively for the center (not the boundaries) of the thermal comfort zone (ASHRAE Standard 55). Conform to the prevailing building code or the ASHRAE standards for thermal (55, 62) comfort and indoor air quality, whichever is more stringent.

- Design for cleanability, maintainability, and simplicity of operation, and then clean, maintain and operate as designed.

- Fully commission the mechanical systems prior to occupancy. Recommission after major renovations or remodeling.

- Periodically check performance and occupant satisfaction. It is a good investment in employee/tenant relationships.

- Avoid expressing negative opinions to occupants about the building, systems, designers, builders, and management.

- Understand the building's liability insurance coverage, and operate within the limits if possible.

New Construction

New building designs should include a comprehensive statement about the HVAC system's performance criteria to support the organization's IAQ goals. The statement should include:

- Suitable design criteria for all appropriate building systems and components.

- A complete description of the HVAC system and its intended operation and performance.

- A commissioning plan including a complete description of the work to be performed during construction.

- Verification procedures for any tests and demonstrations to be performed.

- A complete list of documentation required at the completion of commissioning that can be used as educational tools for operator training of building personnel.

Develop A Policy

An IAQ policy can provide important overall direction and guidance, and can demonstrate part of a good faith effort to provide an environment with acceptable IAQ. This good faith effort

---

Detective Work Can Yield Simple Solutions

Sometimes, a little detective work can result in solving an IAQ problem by finding and eliminating the source. A financial manager at a company complained of frequent headaches at work that disappeared soon after he left the office. After some investigation, it was discovered that the financial manager locked his office every night to secure financial records—and prevented the maintenance staff from running a vacuum over the carpet. As a result, for years the office went uncleaned, causing an unhealthy environment to develop.

Richard Hermans, PE, general manager, facility operations and maintenance for St. Paul, MN Public Schools said at an ASHRAE IAQ seminar that a mechanical approach is not always the correct approach in searching for an IAQ solution to a problem. One case he cited involved a 43,500 sq.ft. school building built in 1924, the subject of allergy-like symptoms and complaints from teachers and students.

Hermans advocates a synergistic, proactive approach to IAQ among occupants and school officials. In this example, the problem was traced to 20-year old carpeting that, coupled with too-high humidity, was a source of mold and airborne spores. A simple carpet cleaning program using bleach and water helped to cure the problem. It was a relatively simple solution to a problem that was serious and yet difficult to determine the cause of—not uncommon in IAQ investigations.

---

is good for employee (or tenant) relations and can help in a court of law.

The Envirosense Consortium, made up of 30 or so manufacturers, consultants and utilities, says "Put it in writing" and suggests that when the building is a rental property, the building owner include tenants in the process.

According to Envirosense: "Even in facilities where no IAQ complaints have been logged, it is essential than an IAQ policy exist in written form and be communicated to the tenants. The plan should include policies, documentation and record keeping, occupant responsibilities, and a complaint-response protocol. Having this information on hand helps tenants understand their responsibilities for protecting the facility's air quality and assures them that building management is equally committed to a healthy environment."

Communicate Effectively

Communication is important, especially concerning IAQ. This is a team effort. Some things the occupants will be able to do for themselves to help. And if there is a problem, it is best to confront it head-on. How a facility manager communicates (or does not communicate) with the building occupants and their means of communication can help to decide whether or not an IAQ concern escalates into an IAQ crisis. Lack of adequate communication between employees and management or occupants and management can leave the impression of an uncaring or ignorant building staff. This can only prolong and/or intensify anxieties about IAQ—office workers have their own means of rapid, broad-based communication, and the office grapevine or water cooler are often it.

Help On Indoor Air And Radon

The State of New York offers help on indoor and radon from the New York State Energy Research and Development Authority in the form of workshops, videotapes and consultation services. For more information, call 518-465-6251, x330.

CARPETING AND OTHER BUILDING MATERIALS

Beware when shopping for carpeting that some can off-gas volatile organic compounds. Ask the seller about this. If possible, air out newly installed carpeting, either before or after installation, but prior to occupancy.

When shopping for carpeting, read labels carefully. The Carpet and Rug Institute (CRI) has developed labels for carpeting that can tell you which types emit lower volatile organic compounds

(VOCs). CRI emphasizes this is not a "warning label" but merely advisory. Carpeting produced in the United States no longer contains formaldehyde; the "new carpet" smell comes from 4PC, which is thought to be harmless and should disappear within a few days of installation. But installation, too, requires caution because of the adhesives used. Low-VOC adhesives should also be used in installation. Carpeting still is the floor covering of choice for most commercial buildings, based on visual image, comfort, warmth, slip resistance and aesthetics. CRI claims carpeting is also economical and easy to maintain.

Table 8-1. Common products and common volatile organic compounds emitted by them.

Product	Common VOCs
Wall coverings	TXIB, Napthalene, Toluene
Floor coverings	Styrene, 2-Ethyl-1-hexanol, Trimethylbenzenes
Paint	Propylene glycol, Butyl propionate, Butanone
Textiles	Formaldehyde, Hexanol, Nonanal
Ceiling tile	Formaldehyde, Acetic acid, Hexanal
Office furniture	Formaldehyde, Acetone, Cyclohexanes

Regular maintenance is the key: Vacuum as often as possible. Carpeting acts as a "sink" for airborne dirt and dust. This is not entirely bad: It keeps it from circulating anyway, and pulls it like a magnet away from your lungs, desktops, etc. and onto a surface we can clean. Deep cleaning should be done at least monthly in lobbies and high-traffic areas—quarterly in corridors and annually in low-use areas such as conference rooms. These are only guidelines, and have to be altered to fit one's own situation.

Some of the same selection care should be used for many types of wallboard and paneling, which contains VOCs as well as formaldehyde. Paint, too, is available in some newer formulations that are less harmful to the environment.

Tools For Schools

The crucial need for better IAQ in schools was met in part by a Tools for Schools program launched by the EPA's Office of Radiation and Indoor Air and sponsored by the National Education Association, the Association of School Business Officials, the National Parent Teacher Association, the Council of American Private Education, the National Education Association and the American Lung Association.

Tools for Schools provides school personnel with a series of checklists they can use to prevent indoor air complaints and problems, rather than having to establish their own procedures after a problem has arisen. There is also information on topics such as control of environmental tobacco smoke, mold and moisture control, and hiring outside assistance.

For more information or to obtain a copy of the program, call (202) 512-1800.

The Maryland State Department of Education also offers several publications helpful to maintaining the indoor air in schools. In Maryland, they are free; mailed elsewhere, there is a small fee. For more information, call 410-767-0096; or write Chief, School Facilities Branch, Maryland State Department of Education, 200 West Baltimore Street, Baltimore, MD 21201-2595.

Sources

Andersson, K., Norlen, U., Ph.D., Fagerlund, I., Hogberg, H., and Larsson, B., "Domestic Indoor Climate in Sweden: Results of a Postal Questionnaire Survey" ASHRAE paper, 1992.

Bernheim, Anthony; and Black, Dr. Marilyn; "Green Architecture: Indoor Air Quality Design is Cost Effective," paper, American Institute of Architects National Convention, May 1995.

Building Air Quality: A Guide for Building Owners and Facility Managers Handbook, Building Owners and Managers Association (BOMA), p. 19.

Environmental Issues in The Workplace II, International Facility Management Association (IFMA).

Fedrizzi, S. Richard, Director of Environmental Marketing for Carrier Corp., and chairman of the U.S. Green Building Council, *Skylines*, January 1995.

Fedrizzi, S. Richard, "Going Green: The Advent of Better Buildings," *ASHRAE Journal*, December 1995.

Gallo, Francis M., "Designing a Proactive Indoor Air Quality Program," *Skylines*, January 1995, p. 18.

Ganick, Nicholas, PE, Ronald V. Gobbell, AIA and Hays, Steve M., PE, CIH, *Indoor Air Quality: Solutions And Strategies*, New York, McGraw-Hill, 1995.

Goldman, Ralph F., Ph.D., "Productivity in The United States: A Question of Capacity or Motivation?" ASHRAE.

Halliwell, Jack L., PE; "IAQ Diagnostics for Building Owners, Managers And Their Consultants," paper at Healthy Buildings Conference, Chicago, sponsored by the National Coalition on Indoor Air Quality (NCIAQ).

Hansen, Shirley, Ph.D., *Managing Indoor Air Quality*, Atlanta: The Fairmont Press, 1991.

Hartkopf, Volker and Loftness, Vivian, Carnegie Mellon University, Pittsburgh, "Innovative Workplaces: Current Trends and Future Prospects" paper presented at the IFMA Conference.

Hedge, A., Burge, P.S., Robertson, A.S., Wilson, S., Wilson and Harris-Bass, J., "Work-Related Illness in Offices: A Proposed Model of The 'Sick Building Syndrome,'" Pergamon Press, Environment International, Vol. 15, No. 1-6, 1989.

Helsing, Knud J., Billings, Charles E., Conde, Jose and Giffin, Ralph, "Cure of a Sick Building: A Case Study" Environment International, Pergamon Press, Vol. 15, No. 1-6, 1989.

Hennessey, John F. III, "How to Solve Indoor Air Quality Problems," *Building Operating Management*, July 1992.

Holcomb, Larry C., Ph.D. and Pedelty, Joe F., "Comparison of Employee Upper Respiratory Absenteeism Costs with Costs Associated with Improved Ventilation," ASHRAE.

Holohan, Dan, "Read Any Old Books Lately?" *PM Engineer*, Business News Publishing Co., February/March 1995.

Indoor Air Quality Update, various issues, Cutter Information Corp., Arlingon, MA.

Int-Hout, Dan, "Total Environmental Quality," ASHRAE.

Indoor Air Quality Manual, Sheet Metal And Air Conditioning Contractors

National Association Inc., 1988, third printing 1990.

Introduction to Indoor Air Quality: A Reference Manual. Washington, DC: U.S. Environmental Protection Agency.

"Keeping HVAC Systems Safe," *Maintenance Technology*, September 1992.

Kirsch, Lawrence, "Liability for Indoor Air Pollution," paper from a symposium at the 198th national meeting of the American Chemical Society, 1989.

Lord, D., Ph.D., "Air Quality in Western Culture: A Short History," ASHRAE.

Lorsch, Harold, Ph.D., PE and Abdou, Ossama, ArchD, "The Impact of the Building Indoor Environment on Occupant Productivity," ASHRAE.

Manko, Joseph, "Investing a Few $$ Can Avert IAQ Legislation," *Econ*, January 1993.

Milam, Joseph A., PE, "A Holistic Approach to Improving Indoor Environmental Quality," Designing Healthy Buildings Conference, The American Institute of Architects.

Mintz, Alan; "IAQ: A Business Opportunity for Contractors," from a talk given at Air Conditioning Contractors of America 26th annual meeting, New Orleans, 1994.

Pomeroy, Christopher D., *Green Building Rating Systems: Recommendations for a United States Rating System*, revised February 17, 1995.

"Productivity And Indoor Environmental Quality Study," National Energy Management Institute, prepared by Dorgan Associates, WI.

"Utility Proves Old Buildings Need Never Die," *Green Building Report*, quarterly publication of the U.S. Green Building Council, January 1995, p. 4.

Wheeler, Arthur E., PE and Olcert, Robert B., Ph.D., CIH, OSD, Indoor Air Quality course, ASHRAE, printed course material.

Chapter 9

Mold: Potential Threats Sprout in a Watery World

It's a big, watery world we live in
and I can't believe it's true
Out of every one in this whole a wide world
Mold found me and you

SHOULD WE BE ALARMED OVER MOLD?

Some people who have suffered from extreme mold infestations would undoubtedly say yes. Many others no doubt have given it a second thought. The current professional prognosis and an accompanying list of concerns lies somewhere in the middle.

"Mold" is defined by Webster's as a superficial often woolly growth produced on damp or decaying organic matter or on living organisms. The words mold and fungi are often used interchangeably. There are more than 20,000 species of mold. Fungi are not plants. Living things are organized for study into large, basic groups called kingdoms. Fungi were listed in the plant kingdom for many years. Then scientists learned that fungi show a closer relation to animals, but are unique and separate life forms. Now, Fungi are placed in their own kingdom.

Most of us don't need a dictionary to tell us what mold is. We see it on the bread that we should have thrown out last week. We see it in the shower stalls we neglected to keep scrupulously clean.

There is nothing in the dictionary definition that is strictly cause for alarm. A moldy piece of bread or cheese is not uncommon; brushing off the mold or cutting it off is all that's required before eating. In fact, some types of cheese (Camembert, for instance) get their distinctive flavor and texture from mold growth. Bathroom shower stalls, curtains and tiles are a frequent source for mold but can be cleaned easily enough with the many products that are on the market for this purpose. Or you can use simple household bleach, diluted with water. Penicillin comes from mold. Mushrooms, both the edible and poisonous varieties, are a form of mold. Most humans are exceedingly tolerant of mold, and it is ubiquitous, for we live in a watery world. It is a part of us. There is a joke that mold is the "state flower" of Florida.

A 1989 NIOSH (National Institute of Occupational Safety and Health) study revealed that only 5% of IAQ complaints were related to microbiological contamination, but more recent studies have shown that there is an association between bioaerosols and increases in IAQ complaints. Inadequate ventilation presented 53% complaints to IAQ, 13% for unknown causes, 4% in building materials, 10% from outside contamination, and 15% from inside contamination.

According to the EPA, there are a very few case reports where toxic molds (those containing certain mycotoxins) inside homes can cause unique or rare medical conditions such as pulmonary hemorrhage or memory loss. These case reports, as stated, are rare. A causal link between the presence of the toxic mold and these conditions is not universally accepted.

According to the Centers for Disease Control in Atlanta: Mold exposure does not always present a health problem indoors. However some people are sensitive to molds. These people may experience symptoms such as nasal stuffiness, eye irritation, or wheezing when exposed to molds. Some people may have more severe reactions to molds. Severe reactions may occur among workers exposed to large amounts of molds in occupational settings, such as farmers working around moldy hay. Severe reactions may include fever and shortness of breath. People with chronic illnesses, such as obstructive lung disease, may develop mold infections in their lungs.

That there is cause for concern is apparent in California Senate Bill 732, the Toxic Mold Protection Act, signed into law in October 2001 making California the first state to develop standards for permissible exposure limits to mold.

The law requires that the California Department of Health

Services establish a mold task force composed of health and medical experts, education and county representatives, and corporate executives to make recommendations for standards in indoor environments. Similar standards for hospitals, nursing homes, and childcare facilities are to be recommended as well. The task force is also responsible for advising the department on standards for the identification and remediation of mold.

The bill also will require landlords and homeowners to disclose the presence of toxic mold when selling buildings.

Recent outbreaks of toxic mold have captured the attention of lawmakers in Sacramento, including state Sen. Deborah Ortiz, D-Sacramento. Ortiz, author of SB 732, said she introduced the bill after employees of local businesses explained how they were sickened by mold at work.

The department must report its progress on developing the limits by July 1, 2003. Owners and landlords will not have to disclose the presence of excessive mold until at least six months after the department adopts the standards. The other mold measure signed by the governor, AB 284, directs the California Research Bureau to study the effects of toxic mold on health.

FIRST, ADD WATER

To have mold, you need water, or at least high humidity. The ability of air to hold water vapor decreases as the air temperature is lowered. If a unit of air contains half of the water vapor it can hold, it is said to be at 50% relative humidity (RH). As the air cools, the relative humidity increases. If the air contains all of the water vapor it can hold, it is at 100% RH, and the water vapor condenses, changing from a gas to a liquid. It is possible to reach 100% RH without changing the amount of water vapor in the air (its "vapor pressure" or "absolute humidity"). All that is required is for the air temperature to drop to the "dew point."

For human health and comfort, relative humidity in a building should be kept between 30-50%. During winter months, when outdoor temperatures are colder, relative humidity should be kept as close to 40% as possible.

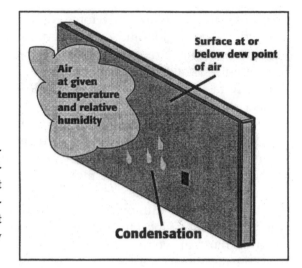

Figure 1-1. Condensation. Air cooled below the dew point will result in condensation on surfaces at or below the dew point.

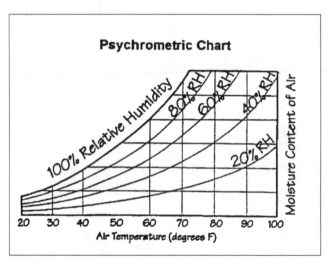

Figure 1-2. Temperature and Moisture. The chart relates temperature and moisture to illustrate the dynamics of moisture control.

Relative humidity and temperature often vary within a room, while the absolute humidity in the room air can usually be assumed to be uniform. Therefore, if one side of the room is warm and the other side cool, the cool side of the room has a higher RH than the warm side.

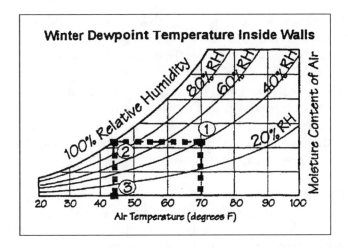

Figure 1-3. Winter Dew Point. Start with the interior temperature, in this example 70°F, and go vertically up the graph to the level of relative humidity, 40% (1). Go left horizontally across the graph to the 100% RH curve (2). At that point, go vertically down the graph to the dew point temperature: 45°F (3).

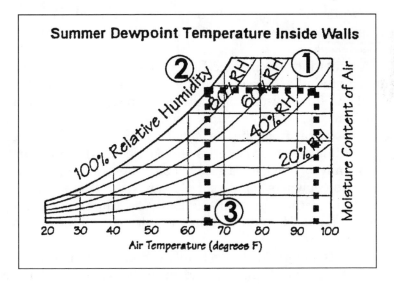

Figure 1-4. Summer Dew Point. In this example, the interior temperature is 95°F (1) and the relative humidity is 40% (2). The dew-point temperature is 65°F (3).

The highest RH in a room is always next to the coldest surface. This is referred to as the first condensing surface, as it will be the location where condensation first occurs, if the relative humidity at the surface reaches 100%.

This is important when trying to understand why mold is growing on one patch of wall or only along the wall-ceiling joint. It is likely that the surface of the wall is cooler than the room air because there is a void in the insulation or because wind is blowing through cracks in the exterior of the building.

Why the Concern over Mold?

Erin Brockovich, whose crusade against Pacific Gas and Electric Co. was made into a hit movie, supported the legislation because her sprawling Agoura Hills home was infested with toxic mold.

So did Steve and Karen Porath of Auburn, who torched their Foresthill, CA, home after they discovered it was infested with toxic mold and other contaminants. The Poraths believe their children, especially 2-year-old Mitchell, became severely ill from exposure to the contamination.

"No family should go through what we went through ever because somebody was allowed to cover it up," Karen Porath told the *Sacramento Bee.*

Another incident that received widespread national attention centered around a 22-room mansion built in Austin, TX. Melinda Ballard and Ron Allison thought it would be their "dream home." For Ballard, a New York public relations executive, it offered a means to retreat from the big city. In 1998, the copper pipes in their building developed a series of water leaks. This problem was subsequently blamed for a nasty mold infestation that sickened the two homeowners and 4-year-old son Reese. The whole family developed breathing problems. Ballard said Allison began coughing up blood and his equilibrium was affected. All three reported suffering memory loss.

Experts determined the family's home was infested with stachybotrys, a black mold found in all 50 states that has been known to cause allergies, asthma and skin rashes. The toxins allegedly got into the home's air conditioning system and were spread. In 1999, Dr. David Straus ordered the family to evacuate their home, leaving at a moment's notice. The family subsequently sued Farmers Insurance Group for what they claimed was its ineptness in handling their claim and failing to adequately remediate the water damage caused by the leaking pipes.

TV star Ed McMahon filed a $20 million lawsuit against his insurance company on his Beverly Hills home. A plumbing pipe burst in July 2001, flooding the den with water. He claims he became seriously ill and was forced to cancel speaking engagements. He was also taking antibiotics after developing a cough and congestion. He moved from the 8,000-sq.-ft. home, and rented another home for $23,000 a month as they resolved the dispute. American Equity Insurance named the suit.

Lack of building maintenance, poor or reduced air filtration, and improper operating procedures also contribute individually to mold and indoor air problems, along with the more visible roof leaks, broken pipes, flooding from rain or fire hoses, sprinkler systems, etc.

The U.S. Supreme Court in 1918 under its United States v. Spearin decided that contractors on a public project, resulting in a defect and specifications furnished by owner. "But if the contractor is bound to build according to plans and specifications prepared by the owner, the contractor will not be responsible for the consequences of defects in the plans and specifications... This responsibility of the owner is not overcome by the usual clauses requiring builders to visit the site, to check the plans, and to inform themselves of the requirements of the work." Many federal and state cases consistently held that responsibility lies with owners when filing suits against the contractors.

"Much of the hysteria surrounding mold can be blamed on insurance companies and the 'mold exclusion' that they now are routinely adding to general liability insurance for subcontractors," said Dave Dolnick, a 25-year veteran in the insurance industry, addressing an ASA (America Subcontractors Association) workshop. "The insurers' responses to mold have been overblown, and are unjustified by either science or the potential liability." Air conditioning and air circulation are the most likely causes of sick building syndrome, not mold.

Facts Table

- In the U.S., about 17 million people have asthma and up to 30% of them are sensitive to molds and fungal allergens. It is estimated that "25% of airways disease and 60% of interstitial lung disease may be associated with moisture in the home or work environment" (HUD, 2001), and where there is moisture, there is most definitely mold.

- All molds have the potential to cause health effects such as allergic reactions, asthma, hypersensitivity pneumonitis, and other immunological effects.

- Inadequate ventilation was found in a NIOSH study to be at the root of 53% of IAQ complaints.

Other Examples of Mold-induced Problems
- A family in Carmel, IN, was forced to evacuate their home after it was contaminated by three types of mold. Ironically, it was built to be a healthy haven for their son, who suffered from asthma. In 2000, Dennis and Debbie Reber, son Peter, and two other children all suffered respiratory complaints, two years after moving into the home. The Rebers surmise the problems began when a tree crashed into their home during a storm, and subsequent repairs failed to keep rain and moisture from entering the home. Every item in the home had to be cleaned, and remediation efforts cost nearly $100,000. The homeowners wondered if it wouldn't be have been cheaper to tear it down and build a new house, especially since there was no guarantee the remediation efforts would be successful.

- Philip Schnepp built a $400,000 dream home in a western Detroit suburb, only to have that dream turn into a nightmare. The 6,800-square-foot home overlooking a trout stream had a problem with mold from sewer backups. Schnepp said his son's eyes were swollen shut from a reaction to the mold. One building, according to an environmental assessment, was full of slimy black mold identified as stachybotrys, aspergillus and cladosporium. Another case of mold infestation gripped several new buildings, but this time it was from a lack of adequate ceiling and roof ventilation.

• Julie and Richard Licon encountered a similar problem in their condominium in Southern California. When they found a black stachybotrys mold on the wood, walls, and floors, they and their six children moved out until the problem could be solved. They moved back in several months later, only to suffer the effects of mold exposure, from nosebleeds to headaches and dizziness. Retesting showed that the cleaning failed to rid their building of mold, that it had manifested itself in the air conditioning system, and was being spread throughout. Once more, they moved out.

Most have a thermometer (there is one on your building thermostat) but few have hygrometers to measure the amount of moisture, or humidistats to control the amount of moisture. In fact, many facility managers are unaware of the humidity level within their buildings, unless it reaches extremes. Too much and you feel stuffy and clammy; metal rusts and wood darkens or rots. Too little and your skin, throat, eyes, and nasal passages feel dry, and you notice that annoying static electricity discharge when you touch something metal.

Temperature is not only easier to control than humidity, it is more visible. Building tenants know when they see 70°F on their thermostat that it is an understandable definition of their comfort. Few know much about proper levels of relative humidity. Just as with indoor air quality, few building tenants are willing to pay more for something they do not truly understand.

Spacious luxury lobbies, better landscaping, better insulated glass, and more wall plugs for computer equipment are easier to grasp than something as invisible as the air we breathe. Over the past several years, there have been a number of young infants (most under 6 months old) in the eastern neighborhoods of Cleveland, who have been coughing up blood due to bleeding in their lungs. Some infants have died and more continue to become ill. This bleeding, a disorder called pulmonary hemorrhage, appears to be caused by something in their environments, most likely toxins produced by an unusual fungus called

The Liability of Facility Managers with
Building Owners under the Law

It was an objective viewpoint from numerous experts in microbiology, medicine, law, engineering and insurance. It called for testing of humidity and mold in residential and small buildings, and meetings for facilities managers who are operating larger buildings. An attorney suggested that everyone learn to the word *vigilant*. "Walk away if you're hesitant," said attorney Malaynne Flehner at an ASHRAE public session.

Really good documentation, building upkeep, and careful maintenance keep you from a session in the courthouse. Stay prepared for a defense for personal injury and property damage, punitive damages, disclosure from your company, and attorney fees. Get educated (include your employer and/or clients).

Dust mites, mold spores, VOCs, and mycotoxins are the new problem in IAQ mold. As long as we have been living in buildings, we've had roof leaks, plumbing problems, and envelope drips and seepage. Buildings aren't drying as quickly as they once did, because they are more moisture sensitive. Allergies and asthma have increased greatly, with unusual indoor mold exposure.

Don't say that a carpet won't inhibit mold; say that a strict antimicrobial has the ability to curtail or restrain moisture. Some IAQ materials enhance products to harbor mold, such as thermally bonded ductwork. Be careful. "Recent health concerns regarding IAQ from ventilation systems have created a need for a duct insulation that will perform without the added potential health risks associated with traditional insulation," according to one manufacturer. It doesn't have the product with the humidity, mold, or even IAQ. It only believes that "the customer wants peace of mind in knowing that the air flowing through their duct system is safe to breathe."

Another manufacturer said technical updates of antimicrobial-coated steel and antimicrobial compound developments affect the types of HVAC applications (ductwork or air handlers) that can use this material. "Many indoor mold, mildew, and fungi problems stem from humidity control issues," said another product manager. "To address this... a wide range of products that provide better temperature and humidity control. Coupling a steel air handler and ductwork, both coated with antimicrobial compound, offers a system with greater protection against these microbes."

For everyone who has to deal with buildings (HVAC contractors, manufacturers, fabricators, specifiers, installers, custodians, facilities managers, building owners, tenants, etc.) some causes for liability are: negligence, breach of contract, breach of expressed warranty, implied warranty, strict liability, and fraud. Insurance claims have gone through the roof in the last two or three years. Increased litigation, loss, and risk make insurers' rates rise.

The insurance aspects of mold issues have been addressed with conferences and conventions such as Basics of Clean Air; the Health Effects of Indoor Mold Exposure; and the Liability of HVAC Industry Members and Building Owners Under the Law.

Personal injuries, property damages, punitive damages, attorney fees—protect yourself again them with maintenance, documentation, and warranties. Be capable and vigilant *now*. And... find a good lawyer!

Stachybotrys chartarum or similar fungi. Pulmonary hemosiderosis, or bleeding in the lungs, is characterized by severe bleeding which can cause coughing up blood or nose bleeds. This is found particularly in infants under 6 months old. Chronic, low-grade bleeding can cause chronic cough and congestion with anemia.

The bleeding is caused by toxins made by an unusual fungus or mold Stachybotrys. When infants breathe in the toxins, the blood vessels in their lungs may become fragile. The weak vessels may be bothered by cigarette smoke or stresses from other illnesses, and start to bleed. You cannot see the toxins in the air; rather, they are carried in the microscopic fungal spores.

Over the past seven years in the Cleveland area there have been 45 cases of pulmonary hemorrhage (PH) in young infants. Sixteen of the infants have died. Thirty-two of the infants have been African American. Most of these cases have occurred within an area of ten contiguous zip codes in the eastern portion of the metropolitan Cleveland. In November/December, 1994, the Centers for Disease Control and Prevention (CDC) lead a case-control investigation on the first ten cases. This study found an epidemiological association of PH in these infants with water-damaged homes containing the toxic fungi, predominantly Stachybotrys. Several lines of evidence suggest that the most likely causal agents are fungal toxins from a fungus called Stachybotrys atra.

This somewhat unusual fungus was found in high quantities in the environments of the affected infants but also to a lesser degree in some of the comparison homes. Stachybotrys requires water-soaked cellulose to grow, and was found in homes where there had been water damage from flooding, plumbing leaks or roof leaks involving wood or paper products (e.g. insulation, gypsum board, ceiling tile). The spores of this fungus contain very potent mycotoxins which appear to be particularly toxic to the rapidly growing lungs of young infants. The linkage of Stachybotrys to PH in infants is on the basis of epidemiological data and has not been conclusively demonstrated. Other factors such as environmental tobacco smoke appear to be important triggers in precipitating overt pulmonary hemorrhage.

More cases continue to occur, a few infants having had only very subtle initial symptoms such as nose bleeds and chest congestion. Concern that there may be a larger number of undetected young infants with this disorder, led to the examination of all infant coroner cases over a 3-year period, 1993-1995. This revealed seven "SIDS" (sudden infant death syndrome) cases with evidence of preexisting major pulmonary bleeding. All but one of these infants had lived in the ten zip code cluster area. This disorder is likely to extend beyond Cleveland since an informal national survey of all pediatric pulmonary centers and continued reporting has identified over 100 similar cases of pulmonary hemorrhage in infants across the country over the last seven years.

WHY IS MOLD A NEW PROBLEM?

Why is mold something new to contend with? Hasn't it always been with us? Building practices since the 1970s are contributing to increased mold growth. "Tighter, warmer, with more furnishings" is how one researcher described this trend. There is reduced air infiltration into our buildings because of added insulation and tighter-fitting doors and windows, which save energy, but which also can lead to a build-up of humidity and other pollutants.

There are other factors which contribute, but some of these are on a more individualized per-building basis. For instance, oversized cooling systems continue to be a sporadic problem, along with roof leaks, leaks around windows, building penetrations, and chimneys and flues, etc.

Even the September 11 terrorist attacks on America have a significance: many building owners and facility managers are considering sealing up their building's ventilation systems to keep them secure from introduction of hazardous materials, such as toxic chemicals or biological agents including anthrax spores. The negative effect on indoor air quality, however, can also sabotage our national well-being.

Bill Coad, president of The American Society of Heating, Refrigerating and Air-Conditioning Engineers Inc. (ASHRAE) said, "Any steps taken that result in a reduction of outdoor air ventilation rates or a change in the manner of providing and treating the ventilation air could seriously change the engineered balance of the interior environment. Such changes can result in many of the manifestations of sick building syndrome causing such maladies as discomfort, eye, nose and throat irritation, headaches, fatigue, lethargy, loss of productivity, upper respiratory symptoms, skin irritation or other sickness." Americans, ASHRAE notes, spend 90% of their time indoors. "Any deterioration of the indoor environment," he added, "would create major health problems."

This isn't as strange as it sounds. *The Air Conditioning, Heating and Refrigeration News* in 1961 reported that many Americans were building bomb shelters in their backyards to prepare for a possible nuclear war between the United States and the Union of Soviet Socialist Republics—without regard to best bomb shelter ventilation practices. The article quoted an expert on the need for proper ventilation for these shelters, complete with manually operated blowers in the event the power went out. Charcoal filters were also recommended "to get rid of radioactive particles, bacteriological agents, and smoke." Presumably, this was to ensure that the occupants of the family bomb shelter, after having survived a nuclear attack, did not subsequently suffocate or have to stew indefinitely in their own juices. "Some type of dehumidification system for the shelter" was recommended as well. IAQ in a bomb shelter was no joke. Two men engaged in a five-day experiment to test psychological and sanitary conditions in a bomb shelter nearly died from asphyxiation on the second day of their experiment. Happily, the *News* reported, "the men were taken to a hospital, given oxygen, and then released."

The potential threat continues today. The Building Owners and Managers Association (BOMA) states: "Release of a toxic chemical into an air handling system is a credible threat" to the nation's stock of commercial office buildings, since toxic chemicals "are readily available in quantities and in forms making them easy to disperse into the air handling system (and) most air intakes are readily accessible." Few such systems are protected by adequate air filters (to do so would be prohibitively expensive, if not impossible) and dispersal of such an agent into the ventilation system would ensure their spread throughout the building within minutes.

BOMA adds: "Our research, at this point, indicates that securing outside air intakes and ensuring basic intruder prevention is the key to success." BOMA advises building owners to safeguard air intakes. "If possible, restrict access to the air intake by ordinary lock and key and intruder alarms at night... if the intake is on the street, perhaps a door could be built to shield it. It is likely that the risk for mischief is less during the day, but if the risk is judged to be significant, then you also might want to install a security camera or perhaps have the area guarded."

Guarding our building air intakes and grilles might be a tough sell to most building owners. But closing those grilles and inlets is a bad idea.

If a chemical or gas is introduced into a building's air intake system:

1) Turn off the air handling system (for an external release) or open up the air handling system to full outside make-up air (for an internal release) and

2) Move occupants away from windows, elevators, courtyards and stairwells and into interior rooms.

Mold isn't limited to homes, certain commercial buildings, or specific regions of the country. It is a growing problem in schools, where younger people who are most susceptible to breathing disorders spend the greater part of their days. According to the U.S. General Accounting Office (GAO 1995; GAO HEHS 95-61), one in five schools in the U.S. has IAQ problems, affecting nearly 10 million students. The American Lung Association estimates that the number of children suffering from asthma is up 42% from 1982. Asthma, the principal cause of school absences, accounts for at least 20% of lost school days in elementary and high schools (Richards 1986.)

Filtration has long been the main line of defense in ensuring quality indoor air. But in the past few years, more attention has been focused on how much genuine good proper filtration can do for any commercial building. Filters must be upgraded from the standard throwaways that sell for a dollar or less and are primarily to protect the HVAC equipment. They will catch baseballs and bugs, but not much more. But filters that are expensive are also poor economy when replaced. Better filters have come onto the market in droves, spurred by advertising, marketing budgets and improved research. Filters won't do anything to lower humidity in a building. They can't control moisture. But they can prove useful in filtering out dirt and mold spores that will lead to mold growth. Some of these filters are disposable and should be changed on a regular basis. Ideally, this means whenever they get dirty. Since most people don't regularly check their filters to see how dirty they are, they should be replaced according to a time-weighted basis, according to the manufacturer's specifications. Some filters are of the permanent type and can be cleaned, either by brushing or with soap and water. Some filters are also available now with features such as activated carbon for odor removal, and at least one even has its own carbon monoxide sensor built in and will sound an alarm when dangerous levels of this deadly gas are detected.

According to Air Quality Sciences Inc. "Air filtration devices must be used to negatively pressurize the mold remediation work area from zones outside the containment barriers. Full contaminants achieve maximum isolation of the work area, but use of tri-chambered decontamination units, and by HEPA filtration of all air exhausted from the containment area."

Fans must be sized accurately in order to be effective. Fans generally begin around 50 cfm for a small bathroom; this may be adequate if the duct run above the fan isn't long or doesn't take too many turns. The farther the fan has to move the air, the more muscle it requires and some small fans just aren't up to the task. Also, many bathrooms now are built larger, with jacuzzis and whirlpools, which contribute much more moisture than a standard bath and shower. A 100 cfm fan may be necessary, but fan size also generally means a noisier motor. Some fans now are being made with quieter motors. If they are too loud, property owners won't want to use them.

On September 4, 2001, students in the Romeo Community Schools District located in an increasingly suburban area 40 miles north of Detroit were to return to class after summer vacation.

Instead of returning to school, however, students in its Washington Elementary School were divided up, assigned to attend classes at a high school cafeteria and a church because of a problem with mold growth. Teachers and some parents had complained before to district officials that the air in the school often made them feel ill and often intensified existing symptoms.

The 44,000-square-foot school was built in 1951 with renovations and additions performed in 1955, 1967 and 1971. As a result, the school had cobbled two heating systems together and an air conditioner for the library.

In 1999, the school was evaluated for poor air quality. "Some people felt that their symptoms got worse while they were in school, so we thought we should check it out," said Paul Soma, Romeo Community Schools' executive director of business affairs. The environmental inspectors found that the air inside the school met all state and federal standards for indoor air quality.

"They did find several areas where we could improve the air quality and that is what we are doing," he said. Later, the situation deteriorated when seven teachers were taken to the hospital and all students evacuated after being exposed to an unidentified contaminant. Environmental experts said it was mold.

The school district called in the local fire and police departments when they first became aware of the emergency. The release of a toxic chemical, on purpose or accidental, or other possible common suspects such as carbon monoxide, was ruled out. After an initial inspection failed to come up with answers, the school system called in Wondermakers Environmental Inc., Kalamazoo, Mich. Wondermakers, Inc. had conducted a survey of the school several months earlier and found that there were several concerns about the quality of the air in the school and made several recommendations. The indoor air quality tests performed by Wondermakers found that there were no immediate indoor air quality problems at the school, and that the mold problem was minimal and did not require closing the school.

"We did not find anything that warranted closing the school. We would have told you if we did," said Michael Pinto, chief executive officer of Wondermakers. Parents questioned Kingsnorth and other administrators about two techniques suggested to contain the mold including encapsulating (painting or sealing) the mold or removing the roof decking. "Removing the roof would be tantamount to demolishing the building. Pulling off the roof decking is an extraordinary thing," Kingsnorth said. Initial tests performed by the two firms showed that there were no mold spores found on black stained ceiling tiles tested in the school.

The district decided to build a new roof over the existing roof, which Kingsnorth and contractors working on the project interpreted as one way to control mold found in the ceiling. The new roof was part of a bond issue approved by voters: $260,000 to replace a leaky roof, with up to $150,000 in updates and repairs to the ventilation systems to improve the school's air quality.

Michael Pinto, CEO of Wondermakers, said that mold counts conducted a few months earlier indicated the air inside the school had 19% of the mold spore count as the air outside of the school. After the ceiling tiles were removed, Wondermakers technicians found that the indoor mold spore count increased to 227% of the mold spore count outdoors. Pinto said that state guidelines suggest that the mold spore count inside a building should not be higher than the outside air.

Wondermakers found two types of mold that indicate poor air quality, including black mold. Pinto said that the mold was a concern and should be cleaned or removed. Pinto recommended a remediation plan that included isolation suits and isolation chambers and that every inch of the school and its contents be thoroughly cleaned of mold by various techniques. He also recommended removal of all materials that were water damaged and encapsulating (painting and sealing) the under decking of the roof and other surfaces where mold was found. The cleanup was expected to last from one to five months. "Part of the goal was to make this as workable as possible," Hill said.

The board's resolutions are in response to public and board fears about mold, particularly Stachybotrys Chartrun or black mold, found at the school during a routine indoor air quality survey conducted at the school last May at the school staff's request.

When the school was evaluated for poor air quality in 1999, Nova Environmental found that the district should:

- Reduce the relative humidity
- Limit the number of plants
- Maintain the heating and air system
- Provide proper exhaust from the men's bathroom in room 100
- Remove water damaged ceiling tiles; and further investigate fungal amplifications sites

Wondermakers Environmental said that none of these recommendations appeared to have been followed when they performed their air quality inspection two years later. An earlier environmental study had a limited scope, Soma said. "It was less than a Band-Aid." The cause of the mold growth was traced to wet ceiling tiles, the result of a roof leak. Romeo Community Schools wound up spending $260,000 to replace the leaky roof and up to $150,000 in updates and repairs to the ventilation systems. Soma said that the district subsequently purchased computer software that will track building maintenance and is instituting a district-wide maintenance program. He said such a program will help to prevent air quality problems in the future because maintenance workers will know when to perform regular upkeep.

Some critics later charged that the entire incident was overblown, and that some of the recommended solutions were "overkill." Replacement of the roof and having cleanup crews wear full body coverage "moon suits" may have been unnecessary. But with the eyes of the media upon this emotional, much-publicized incident, and when the health of the children is at stake, few parents or school officials are willing to take any risks.

Wondermakers and district officials say that the contractor did not follow proper EPA guidelines in cleaning the ceiling tiles, and that workers potentially spread contaminants throughout the school. Wondermakers Environmental made several recommendations in its June 2001 report including:

- Repairing the roof to stop water intrusions
- Maintaining bathroom exhaust fans
- Replacing water-damaged ceiling tiles and exhaust equipment
- Changing the landscape design outside of multi-purpose room to minimize contaminants from outside
- Locating fresh air intakes in the south wind along with installing door vents
- Moving materials away from vents and air conditioning vents in the library
- Cleaning the carpets more frequently

Stachybotrys chartarum (also known by its synonym Stachybotrys atra) is a greenish-black mold. It can grow on material with a high cellulose and low nitrogen content, such as fiberboard, gypsum board, paper, dust, and lint. Growth occurs when there is moisture from water damage, excessive humidity, water leaks, condensation, water infiltration, or flooding. Constant moisture is required for its growth. It is not necessary, however, to determine what type of mold you may have. All molds should be treated the same with respect to potential health risks and removal.

In *Indoor Air Quality Handbook*, John Spengler, Qingyan Chen, and Kumkum Dilwali address the issue of public perception. Public health risks from radon, lead paint, mercury, formaldehyde, asbestos, radiation and chemical waste sites have increased the public perception of other risks, some of them inordinately so. They coin the term "sporophobial" relating to the fear of microbial agents and their infection, allergenic, and toxigenic effects. They also discuss the increasing trend toward litigation: more aggressive action by lawyers, individuals and small groups. Much of this environmental litigation is still in the works so it is unreported, or it has been settled out-of-court. Attorney Mike Diamond offers pointers on how to prevent an indoor environmental lawsuit. Two of the questions defendants must answer in litigation, he says, are whether or not specialists were called in when they became aware of a problem, and were they familiar with the workings of the building's HVAC system and its maintenance.

Many books and guidelines are available now offering how-to advice on keeping a building healthy and safe. Some of it relies on common sense. Not all of it is in agreement, as the science continually evolves, and some of it is downright contradictory although this might also depend on the individual circumstances, surroundings or climate.

For example, it is generally considered to be a good idea to increase the ventilation in a building, to bring in more outside air to dilute any contaminants such as dust particles. But not always. Not if the outdoor air is especially polluted, or if it is extremely humid and you have no way to treat this outside air before it is circulated. In those cases, it might be necessary to improve your filtration first before drawing in more outside air, or recirculate indoor air (assuming it is mixed with fresh air as needed) or de-humidified as it enters the building.

Some indoor air specialists are using IAQ test kits in schools, homes, and commercial facilities to screen for indoor mold contamination. Recent studies have identified Cladosporium, Penicillium, Aspergillus, and Alternaria to be the most common molds found in problem facilities. A summary of recent test kit data shows:

- 42% had the substantial presence of Penicillium and Aspergillus, indicating a current and/or past moisture problem that may warrant further attention;

- 30% had the abundant presence of Cladosporium, Alternaria, and Epicoccum indicating a normal or typical situation, such as settling of these spores from outside air;

- 24% had the abundant presence of Penicillium and Aspergillus, indicating an atypical situation and suggesting a current moisture problem that should have further attention;

- 2% had the abundant presence of Aspergillus versicolor and Aspergillus sydowii indicating a dampness problem that should have further attention;

- 2% had the presence of Stachybotrys and Chaetomium indicating flooding or water damage that should have further attention.

Although the presence of elevated levels and unusual molds can be hazardous, molds and low-level microbial contaminants are common in all environments.

Mold exposure can result in four primary types of illness including infectious disease, allergic reactions, toxic effects, and sick building syndrome or irritation. Molds can produce toxic substances called mycotoxins. Some mycotoxins cling to the surface of mold spores; others may be found within spores. More than 200 mycotoxins have been identified from common molds, and many more remain to be identified, according to the EPA.

Some of the molds that are known to produce mycotoxins are commonly found in moisture-damaged buildings. Although some mycotoxins are well known to affect humans and have been shown to be responsible for human health effects, for many mycotoxins little information is available. Aflatoxin, citrinin, ergot alkaloids, fumonisin, ochratoxin, patulin, sporidesmin, sterigtmatocystin, tenuazonic acid, trichothecenes, and zearalenone can be from peanuts, beans, milk, grains (from corn, barley, oats and rice), coffee, apples, pears, tomatoes... and wallpaper.

In fact, researcher Colin Young, GEOCNO Consultants, San Diego, concluded that biological contamination is a commonly encountered occupational, environmental, and public health issue. Health care and safety professionals should be cognizant of recognizing symptoms consistent with biological exposures. Furthermore, periodic and preventive maintenance of building components so as to avoid water intrusion may be as important as early recognition and treatment of health effects secondary to biological contamination (Mold-Associated Asthma technical paper presented by Colin Young, CIH and Frederick Fung, MD, MS, ASHRAE IAQ Conference 2001). Some common molds or fungi known to produce allergic reactions or sensory irritation include:

Absidia	Curvularia	Mucor	Stachylidium
Agaricus	Dacrymyces	Nigrospora	Stemonitis
Alternaria	Daldinia	Paecilomyces	Stemphylium
Armillaria	Epicoccum	Penicillium	Stereum
Aspergillu	Epidermophyton	Phoma	Tilletia
Aureobasidium	Erysiphe	Pleurotus	Tilletiopsis
Botrytis	Eurotium	Podaxis	Torula
Candida	Fuligo	Polyporus	Trichoderma
Cantharellus	Fusarium	Puccinia	Trichothecium
Chaetomium	Ganoderma	Rhizopus	Urocytis
Chlorophyllum	Gliocladium	Rhodotorula	Ustilago
Cladosporium	Helminthosporium	Saccharomyces	Xylaria
Claviceps	Hypholoma	Serpula	
Coniosporium	Lycogala	Sporobolomyes	
Coprinus	Monilia	Sporotrichum	

Common health concerns from molds include hay-fever like allergic symptoms. Certain individuals with chronic respiratory disease (chronic obstructive pulmonary disorder, asthma) may experience difficulty breathing. Individuals with immune suppression may be at increased risk for infection from molds. If you or your family members have these conditions, a qualified medical clinician should be consulted for diagnosis and treatment. For the most part, one should take routine measures to prevent mold growth. A common-sense approach should be used for any mold contamination existing inside buildings.

Molds such as stachybotrys chartarum and stachybotrus atra are very common in buildings and will grow anywhere indoors where there is moisture. The most common indoor molds are Cladosporium, Penicillium, Aspergillus, and Alternaria. We do not have accurate information about how often Stachybotrys chartarum is found in buildings. While it is less common than other mold species it is not rare.

MOLDS Q & A

Q: How do molds get in the indoor environment and how do they grow?

A: Molds naturally grow in the indoor environment. Mold spores may also enter through open doorways, windows, heating, ventilation, and air conditioning systems. Spores in the air outside also attach themselves to people and animals, making clothing, shoes, and bags convenient vehicles for carrying mold indoors. When mold spores drop on places where there is excessive moisture, such as where leakage may have occurred in roofs, pipes, walls, plant pots, or where there has been flooding, they will grow. Many building materials provide suitable nutrients that encourage mold to grow. Wet cellulose materials, including paper and paper products, cardboard, ceiling tiles, wood, and wood products, are particularly conducive for the growth of some molds. Other materials such as dust, paints, wallpaper, insulation materials, drywall, carpet, fabric, and upholstery commonly support mold growth.

Q: Are there any circumstances where people should vacate or destroy a building because of mold?

A: These decisions have to be made individually. If you believe you are ill because of exposure to mold in a building, you should consult your physician to determine the appropriate action to take.

Q: Who are the people who are most at risk for health problems associated with exposure to mold?

A: People with allergies may be more sensitive to molds. People with immune suppression or underlying lung disease are more susceptible to fungal infections.

Q: How do you know if you have a mold problem?

A: Large mold infestations can usually be seen or smelled.

Q: How do you get the molds out of buildings, including schools, and places of employment?

A: In most cases mold can be removed by a thorough cleaning with bleach and water. If you have an extensive amount of mold and you do not think you can manage the cleanup on your own, you may want to contact a professional who has experience in cleaning mold in buildings and homes.

Q: What should people do if they determine they have Stachybotrys chartarum (Stachybotrys atra) in their buildings?

A: Mold growing in homes and buildings, whether it is Stachybotrys chartarum (Stachybotrys atra) or other molds, indicates that there is a problem with water or moisture.

This is the first problem that needs to be addressed. Mold can be cleaned off surfaces with a weak bleach solution. Mold under carpets typically requires that the carpets be removed. Once mold starts to grow in insulation or wallboard the only way to deal with the problem is by removal and replacement. We do not believe that one needs to take any different precautions with Stachybotrys chartarum (Stachybotrys atra), than with other molds. In areas where flooding has occurred, prompt cleaning of walls and other flood-damaged items with water mixed with chlorine bleach, diluted 10 parts water to 1 part bleach, is necessary to prevent mold growth. Never mix bleach with ammonia. Moldy items should be discarded.

Q: How do you keep mold out of buildings?

A: As part of routine building maintenance, facilities should be inspected for evidence of water damage and visible mold. The conditions causing mold (such as water leaks, condensation, infiltration, or flooding) should be corrected to prevent mold from growing.

Stachybotrys chartarum (Stachybotrys atra) and other molds may cause health symptoms that are nonspecific. At present there is no test that proves an association between Stachybotrys chartarum (Stachybotrys atra) and particular health symptoms. Individuals with persistent symptoms should see their physician. However, if Stachybotrys chartarum (stachybotrys atra) or other molds are found in a building, prudent practice recommends that they be removed. Use the simplest and most expedient method that properly and safely removes mold.

SPECIFIC RECOMMENDATIONS

- Keep humidity level below 50%.
- Use air conditioner or a dehumidifier during humid months.
- Be sure building kitchens have adequate ventilation, including exhaust fans in kitchen and bathrooms.
- Use mold inhibitors which can be added to paints.
- Clean bathroom with mold-killing products.
- Do not carpet bathrooms. Remove and replace flooded carpets.

One air duct manufacturer has developed a coating for the inside of air ducts that is designed to prevent microbial growth. Ductwork is pre-coated at the factory with an antimicrobial to provide a second line of defense in the fight against the growth of bacteria, mold, mildew and fungus in HVAC duct applications. The antimicrobial coating would be appropriate for hospitals and schools, but could also find its way into wider applications.

Building ventilation systems are being viewed more as the "first line of defense" for occupants after the horrific events of September 11, 2001. Not just mold can attack occupants through the ventilation system. "Following the events of September 11, we need to take all steps possible to ensure that America's buildings remain safe, as well as healthy and comfortable," said William Coad, PE, '01 president of ASHRAE.

Coad cautioned well-meaning but overzealous building owners against closing their building air intakes to prevent the introduction of hazardous materials into the ventilating systems. "Any steps taken that result in a reduction of outdoor air ventilation rates or a change in the manner of providing and treating the ventilation air could seriously change the engineered balance of the interior environment," Coad said. "Such changes can result in many of the manifestations of sick building syndrome causing such maladies as discomfort, eye, nose and throat irritation, headaches, fatigue, lethargy, loss of productivity, upper respiratory symptoms, skin irritation or other sickness."

Some alternative solutions are already either at hand or are rapidly being created to protect building ventilation systems and occupants.

Camfil Farr has published an information bulletin on anthrax and the role of air filtration in preventing its spread through HVAC systems.

Titled "Anthrax: Definitions, Questions, Answers & Precautions," the four-page bulletin provides general background information on the disease and discusses the levels of protection afforded by various types of air filtration devices. Additional recommendations for reducing airborne contaminants in buildings are also described.

In addition, the bulletin provides links to web sites offering further information on the topic.

According to this document, "The only true air filtration protection for anthrax is a HEPA filter. To be effective, this must be an air distribution system that takes outside air in through the filter and pushes this filtered air back out through the leak paths in the building—thus the HEPA system creates a slight positive overpressure in the facility." Contact them at 800-333-7320; fax 310-643-9086; or visit *www.camfilfarr.info*.

Isolate Inc., Houston, introduced a line of combination HEPA filters and UVC (ultraviolet) germicidal air filtration units in 2002. Units are sized to provide a minimum of 12 room air changes per hour of 99.99% HEPA filtration efficiency at 0.3 micron particle size while offering an 80% fly-by kill capacity for mold and bacteria, including anthrax spores. A 500 cfm unit has transparent housing panels to reveal the internal unit components. In addition, Isolate has a project involving "negative air" handling filtration and UVC disinfection for mail handling facilities.

AAF International, maker of AmericanAirFilter products, has introduced an "AirShelter" line of products "designed to protect occupants of commercial and public buildings from the risk of airborne contaminants." AirShelter uses HEPA filtration to establish "safe areas" in office buildings, schools, airports, etc. It uses American Air Filter's AstroCel II HEPA filter currently in use in critical health care, government, military and industrial facilities. According to the company, "The combination of positive pressure ventilation (PPV) and HEPA filtration is a proven technology that American Air Filter has used for many years to provide extremely high levels of air purity. This combination virtually eliminates the introduction of airborne pathogens in contained spaces, providing a filtration level greater than 99.97% on 1 to 5 micron particles, such as Anthrax spores."

Meanwhile, **Dual Draw** has introduced a downdraft work station designed to protect mail workers from fumes, powders and other contaminants. It draws these away from the breathing area through 2,500 patented perforations in the table top and vented back, using a 99.97% HEPA filter to clean the air before it is returned to the room. The company says the workstation is quiet at 68 dB and runs on 115V single phase power. Contact them at Dual Draw LLC, 5495 E. 69th Ave., Commerce City, Colo. 80022; 800-977-2125; fax 303-287-0109; *www.dualdraw.com*.

REAL WORLD APPLICATIONS
PROVE SUCCESSFUL

Office Buildings
For every incident of disaster, illness, failed remediation, continuing investigations and lawsuits, there seems to be a success story.

Morning fog and regular rain nine months out of the year in Houston, Texas, created problems for ventilating a 21-story office building. The space houses 33 tenants in 450,508 square feet of space.

For its efforts, Geo Quest Center won an award from the National Air Filtration Association (NAFA). M-81 filters were mounted on three outside air intakes of the building, along with M-81 demister filter frames. Prior to this installation, the intake ducts, pre-filters, and intake side of the outside air handlers became wet whenever there was moisture in the intake air. This moisture prevented the filters from functioning as designed, as well as encouraging the growth of mold, mildew and bacteria.

In addition, the office building upgraded its air handler filtration assemblies, bringing them up to a recommended 65% efficiency rating. A front-loading filter frame configuration was designed by Air Tech Environmental Systems to permit filters to be easily loaded into 8-in. deep steel frames with rubber gasketed seals. A 65% efficient, 6-in. deep rigid box filter with a 2-in. deep 30% efficiency pre-filter was installed in each frame. The new 30% efficiency pre-filters are monitored visually and changed on a preventive maintenance schedule every two months. The new 65% efficient rigid box filters are monitored visually and changed on a p.m. schedule every 12 months. By the end of 2001, all 42 air handlers in the building were converted.

Jails

Correctional facilities have a public perception of being steamy, sultry buildings in summer months, especially those located in the South.

A 14,000-sq.-ft. Bainbridge Probation Substance Abuse Treatment Center in Bainbridge, GA, a minimum security lockup for the Georgia Department of Corrections, was built here with specific IAQ and moisture guidelines in mind.

This sort of project would normally involve the use of conventional DX rooftop air conditioners, and with those units moisture removal would be a challenge.

Current ASHRAE standards of 20 cfm/person of outdoor air for correctional facilities, have made cooling public buildings difficult, especially in highly humid regions. Relying solely on traditional split systems or rooftop units to reduce humidity is both inefficient and sometimes unattainable on hot, humid summer days. The sensible heat ratio of a conventional ac system is at best 0.70 versus the 0.48 required to deliver ideal conditions on this project.

The human element and the building envelope design add to the moisture load as well. Each of the four 2,500-sq.-ft. barrack-style day rooms at Bainbridge, houses up to 25 inmates, and has few windows for security reasons. Because correctional facilities have so few windows, there's no solar effect, which would normally increase the room sensible heat ratio," said Craig Gradick, PE, project manager. "When you combine outdoor humidity with the indoor moisture produced by inmates 24 hours per day in a single room with few windows, controlling humidity can be a prominent problem. Conventional air conditioning will satisfy temperature set points (sensible load) prematurely without running long enough to bring humidity (latent load) to comfortable levels."

Consulting engineer, Jordan & Skala Engineers, Norcross, GA, chose an innovative design involving packaged outdoor air dehumidifiers by Dectron Internationale, Roswell, to supply cooled or heated dehumidified air. Four Dry-O-Tron" DK-30 units and accompanying outdoor condensers satisfy outdoor air requirements by efficiently dehumidifying southeast Georgia's notoriously hot and humid air and using compressor hot gas for reheat when needed.

In order to supply 20 cfm/person of outdoor air, Gradick's air distribution design of 1,850 cfm per day draws 52% outdoor air and 48% return air. This results in 3.8 changes per hour. Each unit is controlled by a Honeywell room thermostat and humidistat.

Specifying a hot gas reheat option on each of the Dectron units also saves on operating expenses. Each unit's hot gas reheat coil uses energy from the refrigeration cycle and does not rely on an added heat source for reheat when in the dehumidification mode. Each unit cools the air to a very dry 48°F dew point, then uses hot gas reheat to warm the supply air to room set point of 75°F and 38% relative humidity. Hot gas bypass provides cooling capacity reduction on off-peak design days. Gradick maintains a neutral building pressure by relieving the outdoor air through fans in each day room's bathroom, which is sized for 2 cfm/sq.-ft. This keeps humid, outdoor air from being pulled inside the building through cracks, openings, etc., adding to the humidity load.

Schools

Unwanted indoor moisture from Florida's steamy outdoor climate damages textbooks, furnishings and school supplies, promoting building deterioration and straining air conditioning systems. Worse, it triggers the growth of microbes that can cause symptoms of illness, trigger allergies and asthma, increase absenteeism, and possibly hinder learning. The school district of Hillsborough County, FL, decided to fight back against humidity when moisture problems triggered staff and student complaints about molds and odors at Tampa Bay Technical High School in 1998. School officials responded to the complaints, cleaning the building and inspecting its ductwork. But problems continued. "When I arrived on the scene, they (school officials) had remediated the mold and were waiting for it to grow back," recalls Ray Patenaude, PE, of St. Petersburg, who specializes in consulting and forensic engineering in connection with indoor air quality issues. "There was a problem in the building, and they had remediated the carpet and ceiling tile, and cleansed and disinfected interior surfaces in the classrooms." Because of the poor IAQ, teachers and students reported medical problems such as upper respiratory system distress and asthmatic reactions, Patenaude said.

Tampa Bay Tech consists of multiple buildings connected by what Floridians call breezeways—essentially a canopy over a sidewalk. Such a setup makes it challenging to control the entry of humid air into classrooms when 1,800 students are entering and leaving every day. The school offers specialized medical, technical, and academic instruction programs. Moisture conditions inside the building, as well as a new ventilation standard (std. 62-1999) established by the American Society of Heating, Refrigerating and Air Conditioning Engineers (ASHRAE) made it necessary to increase the amount of outside air brought into the buildings. Patenaude says the use of electrical equipment to resolve the problem would have been prohibitively expensive, requiring increased power connections and a new transformer. They chose a Munters desiccant dehumidification system regenerated by natural gas. School officials believed the building's IAQ problems stemmed from humidity hovering in the 65-70% range at 95 degrees F. "Florida has the worst ambient average relative humidity of any state in the Union," said Gary Evans, HVAC product sales manager for System Components of Tampa, which sold the desiccant system.

Desiccants are materials with a very high affinity for moisture that pull water vapor from humid air through a process known as sorption. The most common commercial desiccant systems employ wheel technology in which a rotor made of a lightweight honeycomb-shaped matrix impregnated with solid desiccant material rotates between two air streams. As outside supply air passes through one section of the wheel, it comes into contact with the desiccant material, which draws out the moisture.

The wheel rotates slowly into a second opposing reactivation air stream where the desiccant is dried, and the process repeats itself. The desiccant is never "used up" in the process. This is adsorption, which does not change the material in any way, and differs from absorption, in which the substance soaks up moisture and holds onto it, like a sponge.

Active desiccant systems require heat energy to reactivate. Unless supplied by an engine, turbine or other "free" heat source, this is done by a direct-fired gas burner or heat from a steam or hot water coil.

Passive desiccant wheels rely on dry air, which is usually the building exhaust air. Both types of desiccant systems have advantages. Active desiccants dry more deeply and provide better humidity control. Passive desiccants dry more cheaply, with their effectiveness dependent on the dryness of the building's exhaust air.

The use of a 10,000 cfm desiccant dehumidification system dried the air in this case, assisting the air conditioning system in its cooling functions. Desiccant systems work by blowing fresh air through a revolving wheel containing desiccant material that absorbs airborne moisture. Heated air sent through the other side of the wheel regenerates the desiccant to drive out its moisture. The fresh air then circulates through an HVAC system, which cools or heats it, as needed. This system lowered indoor relative humidity to comfortable and healthy levels and kept the mold from returning. Relative humidity indoors now stays in the 50-60% range, dry enough to eliminate any previous mold and dust mite problems.

The desiccant can be a single material, such as silica gel, or a combination, such as dry lithium chloride mixed with zeolites. Other manufacturers also have energy recovery wheels, or enthalpy wheels, to be used in conjunction with their air conditioning equipment to reduce the cost of dehumidifying outside air. Enthalpy is defined as the total amount of heat energy present in a given substance.

Mold, Moisture Problems in Schools

Mold problems are frequently seen in school buildings, particularly but not limited to the southeastern portion of the United States. Schools regularly encounter mold problems because of operating practices that call for shutting off the building cooling and/or ventilation systems during the summer when school is out. A regular practice in many areas is removing mold from books and furnishings just before students return in the fall. Another common practice is to turn the ventilation and/or dehumidification systems off at night, or at the end of the school day, to save energy. Humidity quickly builds up inside the building, with the HVAC system forced to "catch up" just before students return to classes the next day.

"Ventilation air dilutes and removes stale, contaminated air from occupied building zones replacing it with cleaner, conditioned air. Cycling of HVAC systems results in periods of time when contaminants are not being diluted and removed but rising in concentrations," said Charlene Bayer, Georgia Tech Research Institute. She said schools are "far more susceptible" to developing IAQ problems than most other types of facilities, and the children who occupy them are also far more significantly affected than adults. She quoted a U.S. General Accounting Office report that one in five schools in the U.S. has IAQ problems and 36% of the schools studied had "less than adequate" HVAC systems.

Use of desiccant systems to remove moisture is recommended in many areas. Bayer said, "These systems are able to supply continuous ventilation to a space while continuing to control humidity." A study she helped to conduct found that schools had problems with moisture, especially among those that met the recommended ASHRAE ventilation standard of 15 cfm per building occupant—unless they had an active desiccant system. With a desiccant system operating, the maximum CO_2 concentration was about 800 ppm. It rose to about 1,150 when the desiccant system was turned off. "The CO_2 levels throughout the day with the [desiccant system] not operating were much more variable and these variations were dependent on the movement of the students in and out of the room, such as at lunchtime; and the CO_2 concentrations decreased very slowly after the end of the school day. It can be assumed that VOCs and other indoor pollutants would behave in a similar manner to the CO_2, since CO_2 can be used as a surrogate to indicate gaseous pollutant behavior indoors." Maintenance also continues to be an issue in schools. Bayer told of a school custodian who told her he was unable to change the filters in ceiling-located heat pumps because his school district repeatedly turned down his requests for a taller ladder!

In one research project she was involved in "The CO_2 levels resulting from reduced ventilation air clearly demonstrated the need for optimum maintenance practices in the schools."

A frequent source of moisture intrusion in buildings comes from HVAC equipment located on the roof, such as chillers or ventilators. Many of these are protected by sheet metal enclosures which are placed directly onto the roof surface or a roof curb, but the tops of the cabinets lack a roof-like protective coating or surface. Sometimes an improvised temporary vinyl sheet or similar material is placed over them. Leaks that develop can allow moisture to infiltrate into the cabinets, with mold growth taking place right where the air is circulated throughout the building.

Roofs are more likely to fail at equipment curbs, Lawrence Schoen, PE, pointed out, versus flat expanses where the roofing material is continuous. "The weight of rooftop equipment can further depress the roof structure, resulting in ponding in a location where leaks are common." There are other sources of building moisture. "In some cases," according to Schoen, "ducts on the exterior of a building, often the roof, supply air to the interior. If the ducts carry heated or cooled air, there is an insulation system present and it is often sealed with a brushed mastic and sometimes a reflective coating. Our experience is that such coatings remain waterproof for a maximum of five years. The fact that personnel commonly walk on ducts on roofs further compounds this problem. This compromise of the insulation can have several results. In hot and humid climates, condensation can occur on the outside of the duct when it contains cooler air. In cold climates, condensation can occur on the inside of the duct. In a climate with significant precipitation, liquid water, ice and snow can collect on the metal surface of the duct. All these effects can eventually lead to leaks in the duct itself and the resultant entry of moisture into the building."

To prevent this, Schoen advises treating the top and side surfaces of the duct with "an appropriate roofing material and not a brush-on product." He said a rigid insulation board covered by a welded seam roof membrane works well. "Although this costs about three times as much as the alternatives, it performs leak free four to 10 times as long." Ducts that do not require insulation for thermal purposes, he added, should nevertheless have an insulation layer thick enough to cover sharp edges of the metal and an outer roof membrane. Alternatively, the seams in the metal duct may be welded and no insulation applied. In all cases, the top of the duct and insulation system (when present) should be crowned to reduce ponding.

How Do I know if There is Fungus or Mold in Our Facility?

This fungus or mold grows only on wood or paper that have gotten very wet for more than a few days or so. (It does NOT grow on plastic, vinyl, concrete products, or ceramic tiles). If the wood/paper gets wet and is not cleaned up and dried, the fungus may grow and spread. The fungus is black and slimy when wet. It is NOT found in the green mold on bread or the black mold on the shower tiles (but the shower tiles should be kept clean too). If you have had plumbing leaks, roof leaks, flooding in the basement (even if you don't use the basement), or sewer backup in the past year, look for mold or a musty odor.

Common areas for this mold growth include: Water soaked wood, ceiling tiles, wall paneling, unpainted plaster board surfaces, cotton items, cardboard boxes, and stacks of newspapers. If these areas have been very wet, usually for longer than one week, check for mold. After the area dries, the fungus will not continue to grow, but the black dust caused by the fungus can be sucked up by the furnace blower and spread throughout the building. Be sure and check your basement for the black mold. If you do not have access to the basement, ask your landlord for assistance.

Note: not all black mold is Stachybotrys, but moldy buildings are *not* healthy!

Heating Systems

If you have mold, in facilities check to see if there is any way that your forced air furnace can send the mold dust up to the living spaces. Is there ductwork connecting the cold-air returns to your furnace or does your furnace pull air? The latter is the case if you can see the furnace filter face-on (rather than just the edge).

Mold can be found almost anywhere in a commercial building, but especially where a moist environment is likely to be en-

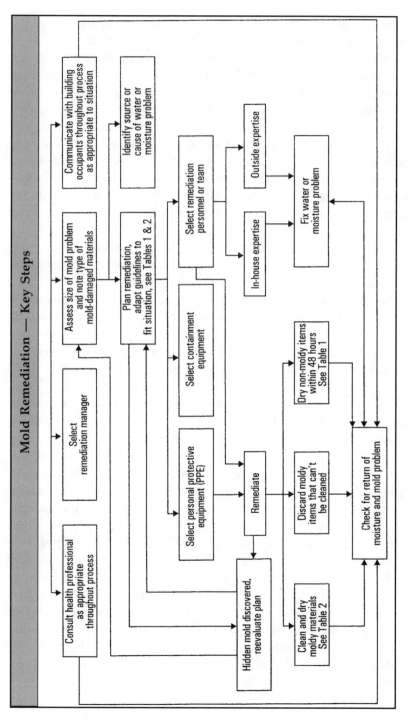

Figure 9-5. Mold Remediation—Key Steps

countered: basements or any part of the building that is underground; kitchens and bathrooms; anywhere there are "leaks" to the outside, such as poorly fitting windows or atriums. But one place you sometimes find mold is inside the air ducts, where it can be invisible to a building's occupants. Air ducts are often round or square bare metal carrying air that is cooler (at least in the cooling season, which varies according to geography) than the surrounding air. This can lead to condensation on the outside of the duct, where it is visible, or on the inside of the ductwork where it can be spread throughout the occupied space. Cooling coils, condensate pans, filters and humidifiers are often HVAC sources for moisture buildup, with the accompanying possibility for mold growth.

The National Air Duct Cleaners Association (NADCA) has published a document addressing mold in air ducts: "Understanding Microbial Contamination in HVAC Systems."

Sampling for microorganisms within the ductwork can be difficult, in fact some refuse to do it all because the results may be meaningless or easily misinterpreted. Some reason that if you see mold, it should be cleaned: it doesn't matter what type of mold it is. However, if a mold can be identified as being especially hazardous, it might require more care in its removal.

Sampling requires some level of expertise, both in the act of gathering the samples themselves and in interpreting the results. Different contaminants require different sampling methods and agents. Also, some duct cleaners may use a sample's results to sell expensive or unnecessary services: finding dust mites, for example, within an air duct system should not be considered cause for alarm. Dust mites are found virtually everywhere that people are found. A visual inspection of the air delivery system is often quicker, easier and more effective. Sampling for microorganisms can be effective if an allergic reaction has occurred. Duct cleaners can visually inspect the inside of the air ducts by cutting holes in the ductwork, which they later seal, or through existing access doors. They also sometimes will use borescopes, video cameras, and other high-tech gear. Once mold has been identified, it must be cleaned but the underlying cause for the growth must also be eliminated or the mold growth will merely reoccur. Normally, the inside of the duct won't promote mold growth.

Sometimes a fiber insulation is used on the interior or exterior of a building's ductwork. This often helps prevent condensation. However, if this insulation becomes damp for any reason, it can act as a sponge to hold in moisture long enough for mold to grow. Nowadays the insulation is often treated with an antimicrobial to prevent such growth. But if the insulation becomes damp it often must be removed, at great cost and inconvenience.

There has been considerable debate over the years on the need and effectiveness of cleaning air ducts. Not too many years ago, this was considered to be a somewhat shady practice in the industry, with many charlatans trying to make a quick buck off of unwary consumers. Many duct cleaners belonged to the "blow and go" category—they had little expertise in what they did; the industry had few or no standards on the work to be done and the equipment was haphazard or homegrown. Over the past 20 years the industry has grown and become accepted. There are professional associations, standards and certification in place now that can ensure a safe, cost-effective job. But the question remains: just how necessary is it to have the air ducts in office building cleaned?

The EPA has waffled, saying the air ducts should be cleaned only when there is a sound need to do so. But industry-wide, there is increasing acceptance for the practice, particularly as standards and certification programs have been put into place.

One duct cleaner equates the air ducts to having one room that is never cleaned. Others say the air you breathe travels through ductwork to reach you, so doesn't it make sense to occasionally clean them? Some say they have found unusual things while cleaning: dead animals, leftover construction debris (frequently the last contractor out has someone sweep the floor) of sawdust and other junk, often sweeping it into the ducts before registers are put into place. More than one duct cleaner claims to have found some construction worker's leftover lunch.

As for mold, mold can indeed grow inside air ducts. Sometimes this happens where humidification is introduced; other times it is found where the cooler air moving through uninsulated metal ductwork meets warmer surrounding air, such as that found in a summertime attic space. Removing mold from ductwork can require some expertise. It is often not enough to insert a vacuum device, which might be fine for stirring up dust and loose dirt particles. Some ductwork mold must actually be scrubbed. Fortunately, there is a host of tools that can be used, most of them developed within the last few years, that will use soft fiber-like whips (so as not to damage the ducts) to scrub the mold and cause it to become airborne enough so that it can then be vacuumed and removed.

Flexible duct and nonmetallic duct, found in some buildings, is not as easily cleaned. Extra care has to be taken so that this often fragile ductwork is not damaged in the cleaning process. Some commercial buildings, such as museums, older buildings built without ductwork, and factories, occasionally use a fabric ductwork that has become more and more popular in the last few years. This ductwork is not only cheaper to manufacture and install than metallic ductwork, but it can be removed, cleaned and rehung much like you would with the curtains or window blinds.

In a guide, "Mold Remediation in Schools and Commercial Buildings," the EPA advises: "Do not run the HVAC system if you know or suspect that it is contaminated with mold. If you suspect that it may be contaminated (it is part of an identified moisture problem, for instance, or there is mold growth near the intake to the system,) consult EPA's guide "Should You Have the Air Ducts in Your Home Cleaned?" before taking further action.

NADCA, which has been instrumental in "cleaning up" this industry, tells consumers on its website (*www.nadca.com*) that "Heating, ventilation, and air conditioning systems have been shown to act as a collection source for a variety of contaminants that have the potential to affect health, such as mold, fungi, bacteria, and very small particles of dust. The removal of such contaminants from the HVAC system and home should be considered as one component in an overall plan to improve indoor air quality." It is certainly out to promote the best interests of its members, but NADCA has also done much to certify its members in safe, effective air duct cleaning practices, and has developed standards that are now written into many commercial bid specifications by building owners and managers.

Also still under debate is the safety and effectiveness of using biocides in mold remediation. Some duct cleaners refuse to use any of these chemicals at all, both because they don't want to assume the liability and because the ultimate effects of these chemicals on building occupants isn't always clear. But others claim they are safe and effective, when used as directed, and may be necessary to thoroughly cleaning the ductwork—that vacuuming alone won't do the job. These must be used with caution, however. However, others say their use is necessary if microbial growth is found, that they are considered to be safe if identified so by the federal EPA and pre-approved for use in air ducts. Some of these chemicals will continue to kill microbes and mold spores long after the ductwork has been cleaned. But the dead mold spores should still be removed.

"It is necessary to clean up mold contamination, not just to kill the mold. Dead mold is still allergenic, and some dead molds are potentially toxic," according to the EPA. Use of a biocide, such as chlorine bleach, is not recommended. It doesn't completely rule out use of biocides by professionals, if done with the proper care and guidelines.

Any chemicals can be considered harmful if misused, however, and not all air duct cleaners can be considered to be properly trained in their use. Some duct cleaners also are offering antimicrobial coatings that remain in place after they have cleaned the ductwork to discourage future mold growth. A few duct manufacturers are even incorporating these coatings inside the ductwork at their point of fabrication as a selling feature. However, others argue that if the source of the moisture is eliminated the use of such antimicrobials may be unnecessary. Some manufacturers are using the biocides to coat filters as a means of killing the microbes before they reach the ductwork. Again, the jury is out on just how safe and effective this is, but there is probably no harm in using the products that have been approved for sale. Duct cleaners will also use biocides to clean drain pans and other portions of the HVAC equipment. Drain pans, used to catch condensate from the air conditioning system, can become breeding grounds for microbes if they become clogged or laden with dirt. A mild bleach solution can be used. Some products have been approved specifically for use in HVAC systems. Beware of some air filters that offer to "freshen" up the air merely by adding a chemical scent that will disguise any problems.

Frequency of air duct cleaning depends on several factors, including:

- Smokers in the facility.

- Pets that shed high amounts of hair and dander brought to pet owners.

- Water contamination or damage to the HVAC system.

- People with allergies or asthma who might benefit from a reduction in the amount of indoor air pollutants in the HVAC system.

- After recent building renovations or remodeling.

Testing for Mold

For a more detailed discussion on the assessment and remediation of Stachybotrys in indoor environments, please refer to The Proceedings of the International Conference held on October 6-7, 1994 in Sarasota Springs, New York, entitled "Fungi and Bacteria in Indoor Environments," pages 201-207, published by the Eastern New York Occupational Health Program. In cases of minor mold contamination, small isolated areas (2 to 10 sq.ft.), testing is usually not necessary.

In cases of more extensive contamination, testing may be necessary. Some private environmental consulting firms may have the ability to conduct commercial buildings assessments and sample for mold identification.

For more help, refer to the section of your yellow pages entitled "Environmental Consultants" to find a company in your area that might be capable of performing these tasks effectively. Ask if the company has experience with mold testing; it is recommended that several price quotes be obtained for field work and analysis.

Consulting firms should be familiar with the American Industrial Hygiene Association (AIHA) document entitled "Field Guide for the Determination of Biological Contaminants in Environmental Samples."

Table 1. Summary of the USEPA BASE Study (Womble et al., 1996, 1999)

Purpose	$	To characterize parameters commonly associated with indoor environmental quality in public and private office buildings
	$	To characterize occupant perceptions of health symptoms and indoor environmental quality in the same buildings
Overview	$	100 public and private office buildings in the continental U.S.
	$	Ten climate regions based on engineering design conditions for summer 2.5% and winter 97.5% dry-bulb temperatures (ASHRAE 1989)
	$	Number of buildings studied in each zone determined proportionally from the populations of the eligible cities in the zone
	$	Buildings selected randomly (except for exclusion of buildings with highly publicized indoor air quality problems) with the intent that problem and non-problem buildings thereby be included in proportion to their existence in the overall population
	$	Data collected during a one-week period according to a consistent protocol
	$	Outdoor samples collected near the source of air for the air handling unit serving the indoor test space
Criteria for indoor study area	$	Occupied by ≥50 employees who could receive a questionnaire
	$	Occupants located on ≤3 floors
	$	Area served by ≤2 air handling units
Descriptive information	$	Building characteristics (e.g., age, construction features, number of floors, floor area, use, occupant number, smoking policy, furnishings, renovations, local point sources, general maintenance)
	$	Operation of the HVAC system
Environmental parameters	$	Airborne culturable fungi and bacteria and total fungal spores
	$	Temperature, relative humidity, carbon monoxide, carbon dioxide, volatile organic compounds, aldehydes, particulate matter, and radon

HOW TO CLEAN UP FUNGAL GROWTH

If you have more than two square feet of mold growth you should seek professional advice on how to perform the clean-up. But the following advice is offered by the EPA:

- The source of the water problem must first be corrected. All roof or plumbing leaks/flooding must be fixed.

- All moldy surfaces should be cleaned with a household bleach (like Clorox) and water mix = 1 cup of bleach mixed in 1 gallon of water. You can add a little dish soap to the bleach water to cut dirt and oil on the wall that can hold mold. With good ventilation, apply the bleach water mix to the surface with a sponge, let it sit for 15 minutes, then thoroughly dry the surface. Be sure to wear a dust mask, rubber gloves and open lots of windows when cleaning with bleach water.

If the area cannot be cleaned (as with wet broken ceiling tiles), is too damaged, or is disposable (like cardboard boxes) discard damaged portions and replace with new.

It may be necessary to do more clean up in the building (carpets, crawl spaces, heating ducts) if you have a bad mold problem.

While Stachybotrys chartarum (atra) occurs widely in North America, it is probably rather uncommon to find it in homes. It requires water soaked cellulose (wood, paper, and cotton products) to grow. When wet it looks black and slimy perhaps with the edges white, and when dry it looks less shiny. It is not the only or the most common black mold to be found in these conditions.

If your clean-up is not simple, i.e. your water damage and mold growth are extensive and/or involve structural materials, contact your city or county health department for assistance in assessing the problem. They can put you in contact with environmental laboratories capable of identifying Stachybotrys, and with abatement contractors familiar with the precautions and other specifics important for extensive clean-up. If you have a large area of mold growth (greater than two square feet or so), seek professional assistance in the clean-up. You can get quite ill if you inhale a large quantity of the fungal dust or get it on your skin.

Fact sheets and other indoor air quality related publications including "Biological Pollutants in Your Home" and "Flood Cleanup: Avoiding Indoor Air Quality Problems" are available from:

Indoor Air Quality Information Clearinghouse
P.O. Box 37133
Washington, D.C. 20013-7133
(800) 438-4318 or (202) 484-1307

There are many ways to clean the air, and many ways to prevent mold, depending in part on the source of the contaminants. Electronic and media air cleaners are getting high marks with many homeowners. Media filters are the simpler and generally less expensive of the two. These are simply filters or filter banks that use a fan to draw in air through their filter(s), removing any airborne dirt particles. Some work better than others, depending on how large the filters are and how good they are at filtering out particles. These require maintenance, including regular replacement although some can also be cleaned.

Electronic air cleaners use electricity to add an electrical charge to the airborne particles, causing them to attach themselves to the electronic plates. Filters can clean the air, but are ineffective against moisture. However, some air filters now are incorporating an antimicrobial coating as a selling feature, the theory being that should they become damp or moist, the coating will discourage mold growth.

A dehumidifier is what you need to remove moisture, or at the very least an effective air conditioner. By effective, we don't mean high capacity. Many building owners, when faced with a commercial building that becomes too warm in the summer, mistakenly ask their HVAC contractor to install a larger air conditioner. Some contractors will oblige; others will take the time to explain that this may not solve their problems. An air conditioner should be sized properly for the footprint and the space it is designed to cool. Square footage is important, but so are the number of windows and their exposure to the sun (east- or west-facing versus north- and south-facing, for example.) Also important are the amount of insulation and the type of construction. Construc-

Table 9-2. Distribution of BASE Buildings by Climate Zone, State, and Season Studied.

			Number of Buildings		
Climate Zone	States	State Total	*i* Summer	*ii* Winter	Total by Climate Zone
A Cool winter,					
Dry and cool-to-moderate	CO	3	3	0	6
or hot summer	NV	3	0	3	
B Cool winter,	IL	3	3	0	23
Damp and cool-to-	MA	3	0	3	
moderate summer	MI	3	0	3	
.	MN	3	0	3	
	NY	6	6	0	
	PA	2	2	0	
	SD	3	0	3	
C Cool winter,	MO	2	2	0	5
Damp and hot summer	NE	3	0	3	
D Moderate winter,	FL*	3	0	3	17
Dry or damp and cool-	GA	3	3	0	
to-moderate	MD	3	3	0	
summer	NC	3	0	3	
	SC	2	0	2	
	TN*	3	3	0	
E Moderate winter,	CA*	3	1	2	6
Dry and hot summer	NM	3	3	0	
F Moderate winter,	AR	3	0	3	13
Damp and hot summer	TN*	3	3	0	
	TX	7	5	2	
G Hot winter,	FL*	4	4	0	7
Dry or damp and cool-to	LA	3	0	3	
moderate or hot summer					
H Hot winter,	AZ	5	2	3	5
Dry or damp and hot summer					
I Moderate winter,	OR	3	3	0	6
Damp and cool-to-	WA	3	0	3	
moderate summer					
J Hot winter,	CA*	12	6	6	12
Damp and cool-to-					
moderate summer					
Overall total	25	100	52	48	100

*States with BASE buildings in more than one climate zone (FL: zones D and G; TN: zones D and F; CA: zones E and J)

tion prior to the 1970s will probably have less insulation and looser-fitting doors and windows, allowing for more infiltration of warm air in the summer. A knowledgeable contractor will use a program called Manual J to compute the correct size for an air conditioner.

What does this have to do with mold? If the air conditioner is oversized, it will tend to quickly cool the room or the space in question, then shut off when it has reached its setpoint. This may drop the temperature, but the unit often doesn't run long enough to remove moisture. A cooler but damp, maybe even clammy, room is the result. This induces mold growth. For an air conditioner to also remove moisture, it has to run long enough, longer than it takes to merely cool the space. The primary task of most air conditioners is to cool the air, not to remove moisture. Fortunately they can do both, if they are properly sized.

Other indoor air contaminants can be addressed: carbon monoxide, for instance, is a colorless, odorless, deadly gas that comes from burning fossil fuels. The non-electric water heater (natural gas propane), non-electric space heaters (propane, kerosene, fuel oil), furnaces (natural gas or heating oil), and automobiles all are potential sources of CO. CO monitors are available now that are as cheap and easy to use as smoke detectors, and are required by some building codes but always should be a standard item.

There are other simple ways to reduce moisture. If you have a building that is too humid, you should cut back on the number of office plants—or grow cactus instead. Much of the water that goes into watering houseplants evaporates into the air. Kitchens and bathrooms should have exhaust fans, especially in bathrooms that don't have windows. You also have to use these fans, letting them run long enough to remove the moisture. Most building codes require exhaust fans.

Use of exhaust fans won't remove moisture if the air outdoors is already saturated. However, even in moist climates it will help on all but the rainiest, dreariest of days. New building codes are coming into effect that require some form of ventilation. These are designed to further improve indoor air quality, but again may or may not have a direct impact on moisture. A building can have excellent air quality, that is, be devoid of dirt and other contaminants, and still be too humid. And one within the optimum humidity range could still have poor air quality.

Radon is often considered one of the most insidious IAQ complaints, although it really has little to do with indoor air. Radon is from a radioactive gas that comes from the ground underneath a shelter. It is odorless and undetectable except by radon detectors. You cannot tell if your facility has high levels of radon without the use of such a detector. Radon isn't reserved for one particular area of the country, although pockets of it have been found, for example, where mining activity has taken place, bringing deeply buried rock closer to the surface. There has been much debate within the medial and scientific community on just what levels of radon are dangerous to homeowners. The remediation with high levels of radon gas is expensive and in some cases impossible. New under construction can be fitted with radon gas removal systems; however, you won't know if these methods are necessary until after the building has been constructed.

Mold Challenges in Manufactured Housing

Manufactured housing contributes problems of its own, and while still a small sector of the overall housing stock it is one that is rising faster than other sectors. Of the two million homes built in the U.S. in 1999, 18.5% were built in factories. In 2000, one in six new single-family homes was built in a factory. Tears and rips in the belly board, which is intended to exclude ground moisture from the interior of the building, are often found, according to one expert. Frequent use of vinyl, plastic and other nonporous wall coverings allows moisture to condense. Manufactured homes are built to HUD Code, Title 24 CFR 3280. This code includes elements of several national building codes. According to researcher Neil Moyer, Florida Energy Center, a study of manufactured homes revealed that "All had a forced air distribution system that was generally considered to be oversized. Most had duct leakage that was significant and caused the building to operate in a negative pressure."

There are even more specific challenges that lie in special use buildings, such as museums. Mike Eissenberg, with the National Park Service, worked on humidity control in Independence Hall, Philadelphia. It was part of a multi-million dollar construction

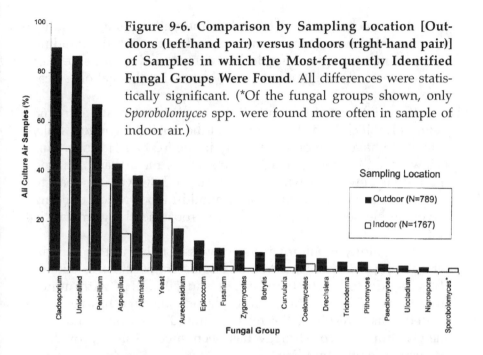

Figure 9-6. Comparison by Sampling Location [Outdoors (left-hand pair) versus Indoors (right-hand pair)] of Samples in which the Most-frequently Identified Fungal Groups Were Found. All differences were statistically significant. (*Of the fungal groups shown, only *Sporobolomyces* spp. were found more often in sample of indoor air.)

Source Removal

Source removal is an option that is used quite frequently today if visible mold growth is present on porous materials. Physical removal is still the best choice for plaster, drywall, ceiling tiles, cellulose insulation, cardboard boxes, and other such materials that harbor visible fungal contamination. In less extreme cases, removing the moisture source and cleaning the surface with a sanitizer/biocidal agent can be effective. If such techniques are employed it is essential that the sanitizing agent comes in full contact with the fungal material for the required amount of time. Whether it is excess humidity, condensation, or moisture intrusion as a result of roof leaks, pipe failures, or subgrade seepage, the water intrusion must be controlled. To err in either one of these critical control activities is to invite recontamination. This is why the ACGIR (American Conference of Governmental Industrial Hygienists) guidelines call for porous materials that have been wet for longer than 48 hours to be removed regardless of whether the leak came from a clean or dirty water source.

Source: An Overview for Safety Professionals
Wonder Makers Environmental Inc.

project that began in 1993 to replace the building's aging utility system. There were three needs that had to be met: the building itself; its many fragile, historical collections and artifacts; and of course the needs of the many people who visit this national treasure annually.

If the interior of such a building is maintained at around 68 degrees F, which is the lower range of the ASHRAE comfort scale, it tends to have an indoor humidity in the 5-20% RH range. This is low enough to induce cracking in some of the artifacts. Raising the indoor dew point with humidification, however, can lead to water condensing from the warm, humid indoor air on building envelope elements such as windows, roof sheathing, metal tie rods.

A humidity monitoring system was installed in this particular building—Hanwell and Vaisala brands. Indoor relative humidity was controlled to between 20 and 70% RH "with minor excursions." Consequently, "little physical damage has been recorded… as a result of the relative humidity swings" and ultimately "this control strategy has been successful in providing protection for the building."

Detecting Mold, MVOCs in Problem Buildings

Early detection and remediation of mold problems is key to minimizing human exposure and extensive building damage. One of the first signs of mold growth is often the presence of damp, musty odors. These odors may occur before any visible mold is present. In addition, it is common for mold to be hidden in inaccessible places such as wall cavities and ceiling plenums.

Mold odors, often characterized as "dirt, mushroom, or musty" result from the release of mold-specific chemicals known as microbial volatile organic compounds or "MVOCs," and research has shown that these chemicals may contribute to adverse human health effects. Last year, Air Quality Sciences Inc., Marietta, GA, announced a new Fingerprint Test for detecting a series of these mold specific VOCs or MVOCs in the air. All air results obtained from schools and commercial environments suspected of having mold contamination have been logged into the National MVOC Databank. Current data trends are summarized below based on all test results:

- Average total VOC concentration is 19,000 ng/m^3;

- 46.2% had 1-Butanol detected at an average concentration of 17,000 ng/m^3;

- 30.8% had 2-Heptanone detected at an average concentration of 1,900 ng/m^3;

- 26.2% had 2-Pentylfuran detected at an average concentration of 1,200 ng/m^3;

- 23.1% had 2-Hexanone detected at an average concentration of 1,200 ng/m^3;

- 15.4% had 2-Methyl-1-propanol detected at an average concentration of 20,000 ng/m^3;

- 7.7% had a-Terpineol detected at an average concentration of 2,800 ng/m^3;

- 6.2% had 3-Methylfuran detected at an average concentration of 1,400 ng/m^3;

- 4.6% had 2-Methylisoborneol detected at an average concentration of 490 ng/m^3;

- 3.1% had 3-Methyl-2-butanol detected at an average concentration of 1,400 ng/m^3.

According to AQS, research has shown that MVOCs easily diffuse through materials and barriers that would normally retard mold spores, and this device provides "an excellent opportunity for finding hidden mold at an early stage."

For more information on the new MVOC Fingerprint Test or to order services call 770-933-0638 or e-mail at *info@aqs.com*.

Design Guide for Humidity Control

"Design Guide for Humidity Control in Commercial Buildings" released in late 2001 was funded jointly by the U.S. Department of Energy and the Gas Technology Institute, and comes from ASHRAE research project 1047-RP. To better control indoor humidity, and to curb any resultant mold growth, this study encourages HVAC designers to:

a)   Focus on adjusting the humidity of any incoming ventilation air

b)   Install dedicated humidity control equipment that operates independently of the cooling and heating equipment

c)   Seal up all duct work and air handlers to avoid pulling untreated outdoor air into building cavities

Microbial growth is a major cause of IAQ problems, according to ASHRAE. In one study of 695 buildings over 10 years, microbial growth accounted for 35% of the IAQ problems encountered. And more than 50% of construction claims against architects, engineers, and construction firms were due to moisture and humidity problems, according to a study of 5,000 claims.

The book asks and answers two basic questions for readers:

1)   What humidity level leads to growth of mold and mildew; and

2)   How high does the humidity have to be held to prevent electrostatic shocks?

Proper use of humidity control through equipment, maintenance and design can avoid these mistakes, according to ASHRAE. Fourteen chapters of the new volume focus on applications designed to help the HVAC designer ask the right questions for a particular building before the design process begins.

Two chapters describe various types of humidification/dehumidification equipment along with their strengths and weaknesses; the use of passive desiccant wheels is also covered. The importance of sealing ducts and the use of humidity sensors is reviewed in a section on instruments and controls.

"Some of the most notorious moisture damage in commercial

buildings is caused by suction from leaking ductwork and air handlers. Duct suction pulls untreated, humid air from outdoors into cool building cavities, where moisture condenses to feed mold and corrode the building structure."

Calculate the moisture load separately from the cooling and ventilation loads. Whenever possible, control the dew point rather than relative humidity.

Calibration of humidity sensors is important, and should be done after installation and prior to commissioning. Most HVAC and controls contractors are not in the habit of doing this, and factory-calibrated sensors may be fine in most cases, with ± 5% RH considered adequate. However, if a building owner has a more defined need, "it is absolutely imperative that all humidity sensors be capable of adjustment after installation, so that their signals are reliable indicators. If sensors are not calibrated in the field before commissioning, it won't really matter what dehumidifier and humidifier capacity is installed —the system will not work as the designer intends."

The EPA offers the following checklist for mold remediation:

Investigate and evaluate moisture and mold problems

- Assess size of moldy area (square feet)

- Consider the possibility of hidden mold

- Clean up small mold problems and fix moisture problems before they become large problems

- Select remediation manager for medium or large size mold problem

- Investigate areas associated with occupant complaints

- Identify source(s) or cause of water or moisture problem(s)

- Note type of water-damaged materials (wallboard, carpet, etc.)

- Check inside air ducts and air handling unit

- Throughout process, consult qualified professional if necessary or desired

Communicate when you remediate

- Establish that the health and safety of building occupants are top priorities

- Demonstrate that the occupants' concerns are understood and taken seriously

- Present clearly the current status of the investigation or remediation efforts

- Identify a person whom building occupants can contact directly to discuss questions and comments about the remediation activities.

Plan remediation

- Adapt or modify remediation guidelines to fit your situation; use professional judgment

- Plan to dry wet, non-moldy materials within 48 hours to prevent mold growth.

- Select cleanup methods for moldy items

- Select Personal Protective Equipment —— protect remediators

- Select containment equipment—protect building occupants

- Select remediation personnel who have the experience and training needed to implement the remediation plan and use Personal Protective Equipment and containment as appropriate

Remediate moisture and mold problems

- Fix moisture problem, implement repair plan and/or maintenance plan

- Dry wet, non-moldy materials within 48 hours to prevent mold growth

- Clean and dry moldy materials

- Discard moldy porous items that can't be cleaned.

Sometimes water and moisture enter a building rather abruptly and visibly. Instead of high humidity or something as often hard to spot as a roof or plumbing leak, moisture comes from a natural disaster like a flood, or even heavy rains. The immediate damage is easy to see and is often contained and cleaned up quickly, but the damage and danger from mold growth can come much later if the necessary precautionary steps aren't taken.

The natural tendency after any disaster is to try to return your life, your residence, or your daily activities to the way they were before, as soon as possible.

Due largely to recent advances in building technologies, it is possible to restore a structure, quickly and economically, having it much more energy efficient than it was before disaster struck.

If you are faced with the task of repairing or rebuilding after a flood, you have several opportunities to make more the structure more energy efficient. By doing so, you will increase your comfort, reduce energy consumption and utility bills, increase your property's value, and save money and energy for years to come.

In normal times, any decision to upgrade or improve the energy performance would have to be based entirely on the energy economics involved.

According to the U.S. Department of Energy, the silver lining in rebuilding is that most equipment or material must be replaced after a flood anyway, so it frequently makes sense to buy the most efficient equipment and materials available.

Repairs to your flood-damaged home can add energy efficiency at the same time you address pressing structural needs, mainly by replacing and upgrading insulation in walls and floors, and checking your foundation for flood damage. Many energy efficiency options are available to you today that may not have been widely available when you built, even if that was only a few years ago.

Cost-effectiveness depends on several factors, including cost of fuel and materials, efficiency levels of the structure and components, and climate. This booklet offers some general tips to improve the efficiency to its shell and equipment: (See the free 60-page booklet titled *Repairing Your Flooded Home* (based for consumers), which is a joint publication of the Red Cross and FEMA. For a free copy, write to FEMA Publications, P.O. Box 70274, Washington, DC 20024, or contact a local Red Cross chapter and ask for ARC 4477.

DOS AND DON'TS ABOUT PERSONAL SAFETY AND ENTERING A BUILDING AFTER A FLOOD

Returning to a building after a flood to survey the damage can be overwhelming. Usually, the first urge is to set upon the task of getting things back to the way they were as quickly as possible. Before any of that can happen, however, you must first verify that the threat of other dangers no longer exists.

DOs:
☐   Walk around the premises and look for structural damage. Note any downed, fallen, or damaged power lines or broken water lines and smell for gas leaks. If power lines or water lines have been damaged, or if you smell gas, notify your utility company immediately.

☐   Wear protective gloves, boots, headgear, goggles, and a respirator (check with your local health department for recommendations on the best type). Flood water contains raw sewage and other contaminants.

☐   Wash hands thoroughly after contact with flood water or flood-contaminated surfaces.

☐   If you are directly exposed to flood water, get a tetanus booster shot if you haven't had one within five years, or if you're not sure when you last had one.

☐   See your doctor immediately if you experience abdominal cramping with nausea, vomiting, or diarrhea.

☐   Contact your county health department about testing your water to be sure it's safe to drink. Have the wells pumped and the water tested before drinking.

❏ Use battery-powered lanterns or flashlights to examine buildings. Never use matches, lighters, or anything else with an open flame as this could cause an explosion if propane, natural gas or other flammable fumes or gases are present.

❏ Keep a battery-powered radio or television handy and tune in frequently for information on where to get medical care, housing, clothing, and food, and on how to help yourself and your community recover.

❏ Report fallen utility lines or submerged utility equipment to the police, fire department, or your utility company. Keep others away until the authorities arrive.

DON'Ts:

❏ Don't eat food that has come in contact with flood water.

❏ Don't use oil or gas lanterns or torches, matches, lighters, or anything else with an open flame that could cause dangerous flammables present to explode.

❏ Don't handle live electrical equipment in wet areas. Always have a professional check and dry electrical equipment before returning it to service. If you're uncertain whether equipment is live, assume it is.

❏ Don't visit the disaster area unnecessarily. Your presence may hamper rescue and other emergency operations.

❏ Don't restart any electrical equipment before having it assessed by a qualified contractor. Submersion of electric or gas heating and cooling equipment can cause system damage. Restarting without proper inspection could result in electrical shock, fire hazard, carbon monoxide poisoning, or explosion.

❏ Don't go near a submerged or overflowing septic tank, or try to service it yourself in any way. If you have a pumped system, disconnect the power to the system before checking it, and don't restart the system without having it professionally checked and serviced.

Now you're ready to begin drying out a building and establishing your plan for rebuilding. The information contained in this section will help you to dry out and decontaminate, sort through the debris, and get ready to rebuild with energy efficiency in mind.

Flooding in a building has several consequences. Materials submerged in flood water can decay, swell, and warp. Electrical equipment and components can become damaged and may cause fires or electrical shock if not replaced after a flood. Wet surfaces encourage mold growth, which discolors surfaces, leads to odor problems, deteriorates building materials, and may cause allergic reactions and other health problems in susceptible individuals. Mud leaves things dirty and the contaminants that may be contained in this mud can pose potential health threats.

Controlling and preventing decay is easier to accomplish and reduces the health risks to occupants if done correctly. Control and prevention of the effects of mold and other contaminants are more difficult to accomplish. However, the stakes are much higher. You can minimize these risks by reducing moisture levels through drying and by decontaminating building surfaces.

Floods are a frequent source of building moisture, and the mold growth they inspire can continue long after the flood waters have abated. After a flood, you must both dry and decontaminate the building.

Either measure alone is not enough.

Because flood water and mud contain sewage, hazardous and toxic materials released upriver, micro-organisms, and other contaminants, it is essential to both dry and decontaminate. Drying without decontamination, or decontamination without drying, are ineffective. Remember too, that all materials and tools you use in the process, such as clothing, wet/dry vacuums, etc., will become contaminated and in need of disinfecting when you are finished.

CAUTION: Contact with flood water may result in severe health risks. Contaminants in flood water can induce cancer, induce birth defects, reduce immune system performance, poison tissues, cause "sick building syndrome," tuberculosis, Legionnaire's disease, aspergillosis, hypersensitivity pneumonia, allergic rhinitis, and viral respiratory infection.

Ventilate the building as soon as possible. Open doors and windows, as well as cabinets, drawers, and closets. Circulate as much air as possible through the building and its cavities, such as walls and attics.

Heat the building as soon as possible. Moisture will move from interior if the indoor temperature is warmer than the outside temperature.

Ventilating and heating are more effective when done together than either is alone.

If your electricity has been safely restored, use portable fans, window air conditioners (set on fan exhaust only), and dehumidifiers to help speed drying. Remember that it is more effective to remove air from the building than to bring air in. Therefore, if you are using a fan to speed drying, aim it toward the outside to move moist air out.

Use a wet-vacuum to remove as much water as possible from the floors and carpets.

CAUTION: Don't use a regular vacuum cleaner on wet carpet. Regular vacuum cleaners are not designed to remove water and mud and attempting this may result in electric shock.

The central fans and blowers of forced-air heating and air-conditioning systems can be disconnected from contaminated ductwork and can be operated for exhausting moist air. Temporary ducts can be connected in such a way that the central blower removes air from the building and discharges it through a temporary duct installed through a window or other opening. This procedure should only be attempted by individuals confident in their ability to rewire electrical systems. If you're unsure how to do this, contact a professional. Also, remember that if the fan or blower motor has been submerged, it will need attention by a professional before it can be safely operated.

Also, while forced-air heating and air-conditioning systems can be used to help dry, because duct work will likely have become wet and contaminated from mud and debris, you will need to clean the ducts before using them. Sheet metal ducts must be hosed out and decontaminated. If liners or insulation are present in sheet metal ducts, remove and discard them. Also, remove and discard fiberglass ducts, as they cannot be properly decontaminated.

CAUTION: Powerful exhaust fans can cause harmful flue gases to be drawn down chimneys and into the living space if the building is closed up tight. If you are using a powerful exhaust fan when heating systems with chimneys are also present, proper venting, such as opening doors and windows, is essential.

Depending on things such as the extent of flooding and the weather, drying your building after a flood could take anywhere from several days to several months, or even longer. You should consult with a professional, such as a local contractor, your county extension agent, or the local building inspector, who may have a moisture meter to test your facility's moisture level.

Since acceptable moisture levels vary in different parts of the country, also consult with these professionals about the recommended levels for your area.

HOW CAN YOU PREVENT RECURRENT MOISTURE PROBLEMS IN THE FUTURE?

Moisture may persist as a problem even after you feel that you have dried out everything that can be dried. Left unattended, excessive moisture can lead to mold, mildew and rot, causing damage that goes beyond the initial mess caused by the flood.

If you find that the relative humidity is higher than 50-60%, you should take steps to reduce it, as this might be the source of problems.

Take care not to aggravate any moisture problems as well. Seek out and reduce the causes of dampness. These few examples might not apply exactly in your case, but they may point you toward potential sources. Good circulation of indoor air prevents trapped pockets of moist air from causing problems.

Your facility should be cleaned from mud and silt immediately to remove any sewage and micro-organisms that may have been deposited on building surfaces by flood water. Removing mud and debris is only the first step. Surfaces that have been cleaned will still be wet and will require time to dry. As these surfaces dry, they will become hosts for mold and other biological growth. Therefore, your facility may have to be decontaminated again once it is dry.

Cleaning or decontamination alone is not sufficient. You will need to do both. To decontaminate surfaces, apply a solution of diluted chlorine bleach on surfaces that have come in contact with flood water or mud.

Experts suggest a solution of 5 to 10% bleach. If you can respond right away, the 5% solution will be enough. The higher concentration is recommended for surfaces that have heavier contamination. In all cases, repeat the treatment at least twice within a 30-minute period.

Be sure the area is well-ventilated, and wear a mask and gloves to protect yourself when doing any cleaning or decontamination work.

New decontamination techniques done by trained professionals are also an option to consider.

Removing Debris

Before installing new materials in your building, remove and dispose of the wet, broken, and unusable materials.

Since flood damage is usually widespread throughout affected areas, there will be a significant impact on local landfills.

Check with your local health department and sanitarian for recommendations on how best to dispose of debris.

Controlling Microbial Growth After a Flood

1. Inventory all flooded areas so that every water damaged area is treated and cleaned.

2. Remove and dispose of wet ceiling tiles and drywall within 24 hours of water contact.

3. Remove and replace all drywall and insulation damaged by water up to 12 inches above the water line. Wicking can cause water to move up several inches above the water level.

4. Dry all wet light fixtures.

5. Water damaged furniture should be replaced or cleaned with a diluted 10% bleach solution. Furniture made of particle board or pressed wafer board should be discarded. Wood furniture can be salvaged by removing microbial growth with a bleach solution. However, check to see whether the solution will damage the furniture finish. Fabrics soaked in standing water should be treated the same as carpets.

6. Leave all cabinets and drawers open to facilitate air flow for drying. All surfaces of cabinets and drawers should be wiped and disinfected with a diluted bleach solution.

7. Remove any essential wet paper from the flooded area to a location where it can be dried, photocopied, and then discarded.

8. If a large amount of files and paperwork cannot be dried within two days, essential files and paperwork may be rinsed with clean water and temporarily frozen until proper drying of the files and paperwork can be completed. Never let paper products become moldy.

9. Immediately remove as much water as possible from wet carpeting using wet vacuums.

10. Upon completion of the wet vacuuming, shampoo the carpet with a 10-percent bleach solution twice within a 30-minute period. Begin shampooing as soon as the wet vacuuming is finished. Before beginning the bleach treatment, conduct a spot test in an inconspicuous area to see if the bleach fades the carpet.

11. If the carpet fades with the bleach solution, then the area must be immediately dried and treated with an alternate biocide. Consult a microbiologist to determine what type of biocide to use, since certain biocides are inhibitors and may not effectively kill microbes.

12. Rinse the carpet with clear water to remove the bleach solution. Take steps to ensure that the carpet is totally dry within 12-24 hours of treatment.

13. Increase air circulation and ventilation if any biocide is used.

14. Air and material testing for microorganisms should be performed immediately after the flood and periodically thereafter by a trained environmental health professional to ensure that no microbial amplification and excessive human exposure occur. Post-cleanup clearance sampling and inspection are necessary to ensure that no excessive concentrations of microbes still exist in the building.

15. Use dehumidifiers and air conditioning/ventilation to speed up the drying process.

    Molds grow on damp materials. The key to mold control is moisture control:

- Lowering the moisture also helps reduce other triggers, such as dust mites and cockroaches.

- Wash mold off hard surfaces and dry completely. Absorbent materials, such as ceiling tiles and carpet with mold, may need to be replaced.

- Fix leaky plumbing or other sources of water.

- Keep drip pans in your air conditioner, refrigerator, and dehumidifier clean and dry.

- Use exhaust fans.

- Maintain low indoor humidity, ideally between 30-50% relative humidity. Humidity levels can be measured by hygrometers which are available.

Sources

IAQ 2001 Moisture, Microbes and Health Effects: Indoor Air Quality and Moisture in Buildings. ASHRAE.

Indoor Air Quality Handbook, 2001, McGraw-Hill.

Builders, Remodelers, and Indoor Air Quality, 1998, Home Builders Institute, National Association of Home Builders.

Airfaqs, Publication of Air Quality Sciences Inc.

Invironment, Chelsea Group Ltd.

Mold Remediation in Schools and Commercial Buildings, EPA, Indoor Environments Div.

Snips, Business News Publishing, June 2003, and various issues.

Appendix I

Trade Associations, Organizations, and Publications

This appendix provides contact information for organizations, trade associations and publications that can offer help and information on IAQ.

GOVERNMENT, ORGANIZATIONS AND TRADE ASSOCIATIONS

Air Conditioning Contractors of America
2800 Shirlington Road, Suite 300
Arlington, VA 22206
(888) 290-2220

Air Conditioning and Refrigeration Institute
4301 Fairfax Drive, Suite 425
Arlington, VA 22203
(703) 524-8800

Air Diffusion Council
11 S. LaSalle Street, #1400
Chicago, IL 60603
(312) 616-0800

American Academy of Allergy & Immunology
61 East Wells Street
Milwaukee, WI 53202
(414) 272-6071

American Conference of Governmental Industrial Hygienists
1330 Kemper Meadow Drive
Cincinnati, OH 45240
(513) 742-2020

American Industrial Hygiene Association
2700 Prosperity Avenue, Suite 250
Fairfax, VA 22031-4319
(703) 849-8888

American Lung Association
1740 Broadway
New York, NY 10019
(212) 315-8700

American Medical Association
515 N. State Street
Chicago, IL 60610
(312) 464-5000

**American Society of Heating, Refrigerating, and
Air Conditioning Engineers (ASHRAE)**
1791 Tullie Circle NE
Atlanta, GA 30329-2305
(404) 636-8400

American Society of Testing Materials
1916 Race Street
Philadelphia, PA 19103
(215) 299-5400

Association of Energy Engineers
4025 Pleasantdale Rd., #420
Atlanta, GA 30340
770-447-5083

Asthma & Allergy Foundation of America
1125 15th Street, Suite 502
Washington, DC 20005
800-7ASTHMA

Building Owners and Managers Association (BOMA)
1201 New York Avenue NW, Suite 300
Washington, DC 20005
(202) 408-2662

Business Council on Indoor Air
2000 L Street NW, Suite 730
Washington, DC 20005
(202) 775-5887

Centers for Disease Control
Mail Stop K-50
4770 Buford Highway, NE
Atlanta, GA 30341-3724
(404) 488-5705

Clean Air Device Manufacturers Association
10461 White Granite Drive, Suite 205
Oakton, VA 22124
(703) 691-4612

Energy-Efficient Building Association
1829 Portland Avenue
Minneapolis, MN 55404-1898
(612) 871-0413

**Heating, Air-Conditioning & Refrigeration
    Distributors International**
1389 Dublin Road
Columbus, OH 43215-1084
(888) 253-2128

Healthy Buildings International
10378 Democracy Lane
Fairfax, VA 22030
(703) 352-0102

Indoor Air Quality Association
10400 Connecticut Ave., #510
Kensington, MD 20895
(301) 962-3805

Indoor Environmental Quality Alliance
24 E. Green Street
Champaign, IL 61820
(800) 556-5595

International District Energy Association
200 19th Street, NW, Ste. 300
Washington, DC 20036
(202) 429-5111

International Facility Management Association (IFMA)
1 E. Greenway Plaza, Ste. 1100
Houston, TX 77046
(713) 629-6753

International Ground Source Heat Pump Association
294 Cordell St., OSU, Room 482
Stillwater, OK 74078
(405) 744-5175

Mechanical Contractors Association of America
1385 Piccard Drive
Rockville, MD 20850
(301) 869-5800

Mechanical Contractors Association of Canada
116 Albert Street, Suite 408
Ottawa, Ontario, K1P 5G3
(613) 232-0492

National Air Duct Cleaners Association
1518 K St. NW #503
Washington, DC 20005
(202) 737-2926

**National Association of Plumbing-
Heating-Cooling Contractors**
P.O. Box 6808
Falls Church, VA 22040
(800) 533-7694

National Air Filtration Association
1518 K St. NW
Washington, DC 20005
(202) 628-5328

National Institute for Occupational Safety and Health
U.S. Department of Health and Human Services
4676 Columbia Parkway (Mail Drop R2)
Cincinnati, OH 45226

North American Insulation Manufacturers Association
44 Canal Center Plaza
Alexandria, VA 22314
(703) 684-0084

National Cancer Institute
Building 31, Room 10A24
9000 Rockville Pike
Bethesda, MD 20892
(800) 4CANCER

National Energy Management Institute
601 N. Fairfax #160
Alexandria, VA 22314
(703) 739-7100

National Environmental Balancing Bureau
1385 Piccard Drive
Rockville, MD 20850
(301) 977-3698
National Radon Hotline
800-SOS-RADON

Office of Conservation and Renewable Energy
1000 Independence Ave., SW, CE-43
Washington DC 20585
(202) 586-9455

Occupational Safety and Health Administration
Office of Information and Consumer Affairs
Room N-3647
200 Constitution Avenue, NW
Washington, DC 20210
(202) 219-8151

Sheet Metal and AC Contractors
National Association
4201 Lafayette Center Drive
Chantilly, VA 22021
(703) 803-2980

The American Subcontractors Association,
the Subcontractor Newsletter
1004 Duke St.
Alexandria, VA 22314-3588
(703) 684-3450
Fax: (703) 836-3482
e-mail: ASAOffice@asa-hp.com

Total Indoor Environmental Quality Coalition
1440 New York Ave. NW
Washington, DC 20005
(202) 638-9015

U.S. Environmental Protection Agency
Indoor Air Quality Information Clearinghouse
P.O. Box 37133
Washington, DC 20013-7133
(800) 438-4318

U.S. Environmental Protection Agency
Indoor Air Division
401 M Street SW
Washington, DC 20460
(202) 233-9030
(202) 260-7751 (Public Information Center)

U.S. Public Health Service
Division of Federal Occupational Health
Office of Environmental Hygiene,
Region III, Room 1310
3535 Market St.
Philadelphia, PA 19104
(215) 596-1888

World Health Organization
Publications Center
49 Sheridan Ave.
Albany, NY 12210

TRADE PUBLICATIONS
AND NEWSLETTERS

Air Conditioning, Heating and Refrigeration News

Airfaqs
A Publication of Air Quality Services, Inc.,
1337 Capital Circle
Marietta, GA 30067
(770) 933-0638
Fax: (770) 933-0641
e-mail: info@aqs.com; www.aqs.com

Engineered Systems

Contracting Business
Penton Publishing
1100 Superior Avenue
Cleveland, OH 44115-2543

Contractor
P.O. Box 5080
Des Plaines, IL 60017-5080

Energy User News

Business News Publishing
(ACHR News, Engineered Systems, Energy News, Snips)
755 W. Big Beaver
Troy, MI 48084

Heating Piping Air Conditioning
Penton Publishing
1100 Superior Avenue
Cleveland, OH 44114-2543

Indoor Air Quality Update
Cutter Information
1100 Massachusetts Avenue
Arlington, MA 02174
Indoor Pollution Law Report

Indoor Comfort News
454 W. Broadway
Glendale, CA 91204

Leader Publications
111 Eighth Avenue
New York, NY 10011
Indoor Pollution News

Buraff Publications
1350 Connecticut Avenue, NW
Suite 1000
Washington DC 20036

Invironment Professional
Chelsea Group Ltd.
One Pierce Place, Suite 245C
Itasca, IL 60143

Pollution Equipment News
Rimbach Publications
8650 Babcock Boulevard
Pittsburgh, PA 15237

Snips

Appendix II

IAQ, HVAC and Energy Terms and Definitions

Absolute Humidity: The weight of water in a given volume of air, often expressed in pounds of water per pound of dry air. See also Relative Humidity.

Absorption: The process whereby a material extracts one or more substances from a liquid or gas, changing physically or chemically in the process. See also adsorption.

Accidental Ventilation: Random air movement into and out of a building caused by accidental pressures.

Activated Alumina: A form of aluminum oxide used as an adsorption material in air filters, more correctly called activated alumina impregnated with potassium permanganate.

Activated Carbon: A form of carbon used as an adsorption media in air filters, often derived from coconut, wood or coal.

Active: Caused by a fan, as in active ventilation or active make-up air supply. See also Passive.

Adsorption: The adhering of a gas onto the surface of a substance without changing chemically; as in an adsorption filter.

Annual Fuel Utilization Efficiency (AFUE): Ratio of annual output energy to annual input energy, including any non-heating season pilot input loss; for gas- or oil-fuel furnaces, it does not include electrical energy.

Air Changes Per Hour (ACH): The number of times the volume of air in a house is replaced with outdoor air in an hour.

Air Exchange Rate: The number of times that outdoor air replaces the volume of air in a building per unit of time, often air changes per hour; or the number of times that a ventilation system replaces the air in a room or area within a building.

Air Cleaning: An IAQ control strategy to remove various airborne particulates and/or gases from the air, generally through the use of particulate filtration, electrostatic precipitation or gas sorption.

Airflow Grid: A device that mounts inside a duct to connect to an air pressure gauge to determine the quantity of air movement.

Air Handler: A cabinet containing a fan used to move air through a system of heating/cooling ducts.

Airtight Construction: A building technique that results in a low leakage rate and high energy efficiency.

Air-To-Air Heat Exchanger (AAHX): A balanced ventilation device capable of transferring heat (and sometimes moisture) between two airstreams.

Allergen: A substance capable of causing an allergic reaction.

Allergic Rhinitis: Inflammation of the mucous membranes in the nose caused by an allergic reaction, also known as hayfever.

Ammonia: NH3 or R-117, a commercial refrigerant.

Antimicrobial: Agent that kills microbial growth; sanitizer, disinfectant or sterilizer.

Asbestos: Fireproof, long mineral fibers once commonly used in insulation and many other industrial and commercial applications.

Aspergillus: Produce potent toxins to humans

Atmospheric Dust Spot Test: A method of testing medium- and high-efficiency air filters.

Axial Fan: A type of ventilating fan having propeller-like blades, often used for window fans.

Backdraft Damper: A device that allows air to only flow in one direction through a duct.

Backdrafting: Reversal of gas flow in an exhaust vent, due to improper HVAC installation procedures or negative pressures indoors.

Balanced Ventilation: A general ventilation strategy that results in a neutral pressure.

Balancing: Adjusting a balanced ventilation system so that the building experiences a neutral pressure.

Benzene: Toxic, colorless, flammable, liquid aromatic hydrocarbon found in solvents and motor vehicle emissions.

Biocide: Technically the same as a pesticide, but sometimes used in HVAC applications such as ductwork to destroy airborne biological contaminants.

Biological Pollutants: Living or organic air pollutants, such as mold and pollen.

Blower Door: A device used to pressurize or depressurize a house to evaluate its efficiency.

British Thermal Unit (Btu): The amount of heat required to raise the temperature of 1 pound of water 1°F.

Building Automation: Any of a number of control systems used to monitor building energy and/or HVAC, lighting or security systems.

Building Envelope: The exterior portion of a structure, usually insulated, that surrounds the conditioned space.

Building-Related Illness: Diagnosable illness whose symptoms can be identified and whose cause can be directly attributed to a source within a building.

Bypass: A pathway through a building, usually hidden within the structure, through which air moves without directly causing an air exchange in the living space.

Capital Cost: The installed cost of equipment.

Carbon Dioxide (CO_2): A natural colorless, odorless gas released in exhaled breath, or by combustion processes.

Carbon Dioxide Sensor: A device that senses CO_2 concentration in the air, often used to control a ventilation system.

Carbon Monoxide (CO): A colorless, odorless, highly toxic gas released during the incomplete combustion of carbon-based fuels.

Carbon Monoxide Detector: A device that senses carbon monoxide in the air and sounds an alarm when the concentration reaches a certain point.

Carcinogen: A substance or compound that can or is known to produce cancer.

Centrifugal Fan: A ventilating fan with blades resembling a squirrel cage, often found in furnaces.

Central Exhaust Ventilation: A general ventilation strategy that uses an exhaust fan to remove air from a building. An equal volume of makeup air enters either by way of through-the-wall vents, through a fresh air duct connected to a forced air heating/cooling system, or through other entry areas.

Central Supply Ventilation: A general ventilation strategy that uses either a supply ventilation fan or a forced air heating/cooling fan to draw air into a building.

Central Ventilation: Any general ventilation strategy—i.e., balanced, central supply or central exhaust.

Chiller: Any piece of equipment used to reduce temperature, but most often used to refer to commercial rooftop type units as part of a chilled water system.

Chlorofluorocarbon (CFC): An ozone-depleting refrigerant. No longer being manufactured or distributed in the United States.

Circulating Air: Air that moves from room to room or area to area within a building.

Clean Room: A specially designed conditioned space in which air quality is strictly controlled.

Combustion Air: Air that enters a house specifically to be used for combustion purposes.

Combustion By-products: Gases and particulates released when a fuel is burned.

Commissioning: Process of testing and adjusting a building HVAC and other systems prior to startup or occupancy to assure functions and performance.

Compressor: A mechanical pump used in a refrigeration or air conditioning system to raise the pressure of a refrigerant and thereby its boiling temperature.

Condensation: The changing of a gas or vapor into a liquid, accompanied by the release of heat.

Conditioned Space: The interior surface area of a building that is kept heated or cooled.

Condensing Unit: Part of an AC/R system that takes low temperature/pressure refrigerant and converts it into a liquid to absorb heat.

Constant Volume (CV): Also called constant air volume (CAV), CV designates an air handling system that provides a constant air flow while varying the temperature according to demand.

Continuous Use: A designation that a fan is capable of running for indefinite periods of time, not the same as continuous duty.

Controlled Pressure: An air pressure difference between the indoors and the outdoors, induced by ventilation equipment.

Controlled Ventilation: Purposeful air movement into and out of a building, usually caused by a fan moving air through deliberate openings.

Convective Air Currents: The movement of air resulting from heat transference through the air by convection.

Cubic Feet Per Minute (cfm): An expression of a fan's capacity. Also a measure of airflow through a space.

Damper: A device, often motorized or manually adjustable, used to vary or control the airflow in a duct.

Decibel (dB): Units used to measure sound intensity. Two decibels are 10 times as loud as 1 decibel.

Degree Day: A unit of measurement used to estimate fuel consumption and heating or cooling costs, based on temperature and time.

Dehumidification: Removing water vapor from the air through chemical or physical means, such as cooling below the dew point level.

Dehumidistat: A control device that can be used to activate a control system as the relative humidity rises.

Demand-controlled Ventilation: The process of automatically supplying air to, and removing air from, a building only when it is required by occupants.

Depressurization: When the air pressure is less than the atmospheric pressure outside.

Desiccant: A substance that absorbs moisture. A desiccant unit is a dehumidification system.

Dew Point: The temperature at which air is saturated with moisture (100% relative humidity).

Diffuser: A grille or cover designed to direct airflow in a pattern into a room for optimum air mixing.

Diffusion: The migration of molecules of a gas or a vapor (or a liquid) from an area of high concentration to an area of low concentration.

Diffusion Barrier or Retarder: A material that slows down the amount of diffusion through a solid material, such as humidity through an exterior wall.

Dilution: The mixing of fresh air into stale air to reduce the concentration of pollutants.

Dilution Air: Air that mixes with combustion by products prior to their being expelled from a vent stack or chimney.

Displacement Ventilation: A controlled method of moving air through a room, generally only used in specialized applications.

DOP Smoke Penetration Test: A method of testing high efficiency (HEPA) air filters.

Duct: A pipe or conduit for moving air, made from a variety of materials and in a variety of shapes.

Ductboard: A semi-rigid fiberglass material with an aluminum foil facing on one side used in ductwork.

Economizer: System of controls or dampers that mix outdoor air with air in a conditioned space.

Energy Efficiency Ratio (EER): A measure of efficiency used to rate equipment.

Electret: A plastic material that carries a permanent static charge, used in some air filters.

Electrically Commutated Motor (ECM): A type of variable speed motor that utilizes electricity efficiently, especially at low speeds.

Electromagnetic Radiation (EMR): The electrical and magnetic energy surrounding electrical wires and appliances.

Electronic Air Cleaner: Devices that draw in room air, clean it and discharge it; many produce ozone as a by-product.

Electrostatic Air Filter: An air filter composed of plastic materials that capture particulate pollutants using static electricity.

Electrostatic Precipitator: An air filter that uses an electrical charge to cause particles to stick to a series of metal plates for their removal from the air.

Energy-Recovery Ventilator (ERV): A heat recovery ventilator that recovers both latent heat and sensible heat.
Enthalpy: The total amount of heat contained in air, the sum of the sensible heat and the latent heat.

Environmental Tobacco Smoke (ETS): Tobacco smoke as it relates to nonsmokers or building occupants.

Evaporation: The changing of a liquid to a gas or vapor, requiring a change in temperature.

Evaporative Cooling: A system using the absorption properties of a liquid, usually water, to cool the air it comes in contact with.

Exfiltration: Air leaving a building through holes or seams in the structure.

Exhaust Air: The air leaving a building through its ventilation system.

Exhaust Grille: A grille air moves through to leave a room.

Expansion Valve: A refrigerant metering device with a pressure or temperature-controlled orifice.

Extended-surface Filter: An air filter with a large amount of surface area but little resistance to airflow.

Exchange Rate: The rapidity at which indoor air is replaced with outdoor air.

Fan: An electrically powered device that moves air.

Fenestration: Light-transmitting envelope components, such as windows or skylights.

Filtration: The process of removing pollutants from air.

Flame Rollout: Flames pushing or rolling out of a combustion chamber, usually the result of backdrafting.

Flat-plate Core: A metal, plastic, or treated paper device that is used in HRVs to transfer heat (and sometimes moisture) between two airstreams.

Formaldehyde: A gas often released from common building materials, consisting of a carbon atom, 2 hydrogen atoms, and 1 oxygen atom. Carcinogenic to humans.

Fresh Air Duct: A duct through which air travels from outdoors to indoors, often used with a forced air heating/cooling system.

Fungi: Any of a group of parasitic lower plants lacking chlorophyll, including mold and mildew.

Grille: An often decorative covering on a wall or ceiling which conveys air moves to or from a conditioned space.

Heat Pipe: A sealed tube partially filled with a refrigerant, used in some HRVs to transfer heat between two airstreams.

Heat Pump: A device using a refrigerant pumped through a system to add heated or cooled air to a building or space.

Heating Seasonal Performance Factor (HSPF): Rating method for total heating output of a heat pump during normal annual use.

Heat Recovery Ventilator (HRV): A ventilation device capable of transferring heat (and sometimes moisture) between two airstreams, or between water and air.

HEPA Filter: For high-efficiency particulate air. A very high efficiency air filter, often used in hospitals and laboratories and in air duct cleaning devices.

Humidifier: A device used to add moisture to the air.

Humidifier Fever: A respiratory illness causes by exposure to toxins from microbials found in humidifiers and/or air conditioners.

Humidistat: A device used to control the humidity in a space.

Humidity: The amount of moisture in a space, whether absolute or relative.

Hydrochlorofluorocarbon (HCFC): Refrigerant slated for phase-out but not as harmful to ozone layer as CFCs.

Hygrometer: A device that senses relative humidity.

Hypersensitivity Pneumonitis: A group of respiratory diseases that cause inflammation of the lung, often caused by inhaling organic dust including molds.

Inches Water Gauge (in. WG or " WG): A unit of pressure measurement. About 250 Pa. equals 1" WG.

Induced Draft: A system that uses a fan to exhaust air from a combustion appliance.

In-line Fan: A ventilating fan made to attach a duct at each end, often called a tube fan.

Insulation: A substance or material used to control the flow of temperature.

Ionizer: A device that generates ions (usually negatively charged), sometimes for the purpose of cleaning the air.

Kilowatt-hour (kWh): A unit of measuring electrical energy equal to 1,000 watts of power consumed over an hour.

Latent Heat: The amount of heat required to be removed from air to change water vapor from a gas to a liquid.

Legionnaires' Disease: An often fatal disease produced by the bacterium *Legionella pneumophila*, sometimes found in HVAC or plumbing systems, particularly those not properly maintained.

Local-exhaust Ventilation: The removal of localized air pollutants, often using a dedicated fan or fume hood.

Magnahelic Gauge: A device used to measure pressure.

Make-up Air: Outdoor air that enters a building to take the place of air that has been exhausted.

Manometer: A gauge for measuring air pressure differences.

Mechanical Ventilation: Supplying and exhausting air to or from a building using a fan or other artificial means.

Media Filter: A filter that relies on a fibrous material to remove particles from the air.

Medium-efficiency Filter: A particulate filter, usually in the 30-40 percent efficiency range when measured on the atmospheric dust spot test.

Micron: A unit used to measure the size of air pollutants or particles; a millionth of a meter.

Mixed-gas Sensor: A device that senses gases in the air.

Modulating Damper: A motorized damper that can be opened and closed a variable amount.

Mold: Mass composed of the spore-bearing mycelia.

Motion Sensor: A control device that senses movement in a room to operate a ventilation, security or lighting system.

Multiple Chemical Sensitivity (MCS): A term used to describe a condition involving hypersensitivities to chemical contaminants, but still not widely or officially recognized in the medical community. Also known as environmental illness.

Mutagen: A substance that causes changes in chromosomes or genes.

Mycotoxin: Mold known to produce potent toxins.

Negative Ions: Negatively charged atoms or groups of atoms.

Negative Pressure: See depressurization.

Net Free Area (nfa): The amount of unobstructed open area of a grille, inlet, or outlet.

Neurotoxin: A chemical or substance toxic to the human nervous system.

Neutral Pressure: Air pressure inside a building that is equal to the atmospheric pressure outside.

Occupied Space: The part of a building normally accessible to occupants, viewed horizontally and vertically, also called the occupied zone.

Out-gassing: Also called off-gassing, the release of volatile gases from a solid material, such as formaldehyde from wallboard.

Ozone: A highly reactive gas consisting of three oxygen atoms; can be poisonous near the ground but is a necessary component of the upper atmosphere as a means of filtering out harmful ultraviolet rays from the sun.

Ozone Generator: An air cleaning device that produces ozone as a means of reducing pollution or cleaning the air.

Partial-bypass Filter: A filter that removes some contaminants from the air, but allows some air to pass through particulate without contacting the filter, usually to minimize the resistance to airflow.

Particulates: Solid (or liquid) air pollutants, as opposed to gases.

Particulate Filter: An air filter designed to remove particulates from the air.

Parts Per Million (ppm): A small unit of measurement, often used to express the concentration of pollutants in the air.

Pascal (Pa.): A small unit of pressure measurement, useful in diagnosing houses. One pound per square inch equals about 7,000 Pascals.

Passive Ventilation: Not caused by a fan or other mechanically induced means.

Pass-through: An opening between two rooms through which air can move when pressure imbalances occur between the rooms.

Permanent Split-capacitor Motor: A type of electric motor characterized by a main winding and an auxiliary winding wired in series with a capacitor.

Permeable Wall: An insulated part of a structure through which outdoor air can pass from the outdoors to the indoors and is tempered in the process.

Permissible Exposure Limits (PELS): Standards for chemical exposure set by OSHA.

Pesticides: Chemical compounds formulated to kill living creatures.

Picocurie Per Liter (pC/l): A unit of measuring the concentration of radon in the air.

Plenum: Air compartment connected to ducts or ductwork.

Pneumatic: Operated by air pressure, as in a control system.

Pontiac Fever: A respiratory illness caused by a form of *L. pneumophila*. The illness last from three to five days, is not related to pneumonia, and has milder, flu-like symptoms.

Pounds Per Square Inch (psi): A unit of pressure measurement.

Pressurization: When the air pressure inside a house is greater than the atmospheric pressure outside the house.

Psychrometry: The science of measuring moisture mixed with air.

Psychrometric Chart: A graph used by engineers to determine the moisture content of air at different temperatures.

Radon: A naturally occurring radioactive gas, often released from soil and rocks.

Radon Daughters: The natural radioactive decay products of radon, some of which release harmful radiation.

Recirculating Range Hood: A non-ducted range hood, one that is not exhausted to the outdoors.

Refrigerant: Any liquid or gas used to transfer heat, usually designated by the letter R, such as R-12 or R-22.

Relative Humidity (RH): The amount of moisture in air compared to the maximum amount of moisture that air at that temperature can contain, usually expressed as a percentage.

Return Air: Air from the living space that enters a furnace or air conditioner to be conditioned.

Rotary Core: A slowly spinning plastic wheel that is used in some HRVs to transfer heat and moisture between two airstreams.

Sealed Combustion: A system used in high-efficiency furnaces, boilers and water heaters that draws combustion air directly into a combustion chamber from outdoors rather than from room air.

Seasonal Energy Efficiency Ratio (SEER): A method of rating air conditioner cooling efficiency.

Sensible Heat: The heat involved to raise or lower the temperature of air, not including any heat required to cause water vapor to change state from gas to liquid or liquid to gas.

Sick Building Syndrome (SBS): Used to describe a set of ailments due to environmental factors affecting the occupants in a particular building.

Short Circuiting: When fresh air moves quickly through a space without time for adequate mixing or dispersal with ambient air. Also used to describe when exhaust air flows unimpeded back into supply air.

Sone: A linear unit of sound measurement used to measure sound levels. Two sones are twice as loud as one sone.

Source Control: The principle of removing sources of pollution or limiting the effects of possible pollutants in the occupied space.

Spillage: A situation where some harmful combustion by-products can spill into the living space, often caused by faulty installation or equipment, or negative internal building pressures.

Spot Ventilation: Removing airborne pollutants at their source, say a cook stove or copier machine, often with a dedicated fan or fume hood.

Stachybotrys atra: Potent toxins under certain circumstances.

Static Pressure: The amount of pressure exerted against the walls of a duct or airway, created by the friction and impact of air as it moves.

Static Pressure Drop: The change in pressure resulting from resistance to airflow.

Supply Air: The air used in the combustion process of a furnace or to supply an air conditioner.

Supply Grille: A grille through which fresh air enters a room.

Synergy: The interaction of two substances whose combined effect is greater than the sum of the individual effects.

Thermostat: A controller for temperature.

Thermostatic Expansion Valve (TXV): A valve that controls the flow of refrigerant.

Tracer Gas: A gas used to evaluate air exchange or indicate air flow in a building.

Unit Cooler/Unit Heater: A direct-cooling or direct-heating, factory made assembly which includes a cooling or heating element, fan, motor and directional outlet.

Unitary System: A room unit to cool the air, using either air or water.

Vapor: The gaseous form of a substance that is normally a liquid at room temperature.

Vapor Pressure: The small pressure exerted by vapor molecules in a mixture of gases.

Variable Air Volume (VAV): A common air handling system that conditions the air to a constant temperature and varies the amount of airflow to ensure thermal comfort.

Ventilation: The process of supplying air to, or removing air from, a building or occupied space.

Volatile Organic Compounds (VOCs): Hundreds of different molecular compounds that contain carbon and easily evaporate. Many VOCs are released from common building materials and found as contaminants in indoor air.

Water Gauge (WG): A device used to measure pressure differences in inches of water.

Weight Arrestance Test: A method of testing low efficiency air filters.

Appendix III

Manufacturers, Suppliers and Service Providers

The directory that follows is a representative but by no means definitive list of those in the industry who are able to provide products or information on subjects in this text. Those companies that wish to be added to this directory should write to the publisher for consideration.

AAF McQuay International
13600 Industrial Park Boulevard
Minneapolis, MN 55440
Tel (612) 553-5330
Fax (612) 553-1667
Unit ventilators, water source heat pumps, air filters

Abatement Technologies Inc.
2220 Northmont Pkwy, Ste. 100
Duluth, GA 30096-5895
Tel (800) 634-9091; (770) 339-3600
Fax (770) 339-2600
Duct cleaning equipment

Aegis Environmental Management
3106 Swede Avenue, Suite 1
Midland, MI 48642-3843
Tel (517) 832-8180
Fax (517) 832-7572
Antimicrobial technology, specific IAQ building programs

Air Diffusion Products Inc.
801 Pressley Road
Charlotte, NC 28217
Tel (704) 527-0763
Fax (704) 527-9153
Air diffusers

Air-Kontrol Inc.
221 Pearson Street
Batesville, MS 38606
Tel (601) 563-0464
Fax (601) 563-0464
Air filters

Airguard Industries Inc.
3807 Bishop Lane
P.O. Box 32578
Louisville, KY 40232
Tel (502) 969-2304
Fax (502) 961-0930

American Sensors, Inc.
30 Alden Road, Unit 4
Markham, ON, Canada L2R 2S I
Tel (416) 477-3320
Carbon monoxide detectors

Aston Industries, Inc.
P.O. Box 220
St. Leonard d'Aston, PQ, Canada JOC IM0
Tel (819) 399-2175
Heat recovery ventilators, central-exhaust systems

Air King Ltd.
110 Glidden Road
Brampton, ON, Canada L6T 213
Tel (905) 456-2033
Thermostatic fresh air control

Air Kontrol
221 Pearson
Batesville, MS 38606
Tel (800) 647-6192
Electrostatic air filters

AirPro, Inc.
8541 Meredith Avenue
Omaha, NE 68134
Tel (402) 572-0404
Pressure controllers and gauges

Airxchange
401 VFW Drive
Rockland, MA 02370
Tel (617) 871-4816
Commercial-sized HRVs

Alnor Instrument Co.
7555 N. Linder Ave.
Skokie, IL 60077
Tel (847) 677-3500; (800) 424-7427
Fax (847) 677-3539
Temperature measuring, air velocity intstuments

Barneby & Sutcliffe
P.O. Box 2526
Columbus, OH 43216-2526
Tel (614) 258-9501
Adsorption filters

Biltwel Duct Cleaning Equip. Mfg.
16327 130th Ave., #4
Edmonton, AB T5V IK5 Canada
Tel (800) 547-5210
Fax (780) 447-2061
Manufacturer of duct cleaning equipment

Bio-Cide International Inc.
2845 Broce Drive
P.O. Box 722170
Norman, OK 73070-8644
Tel (405) 329-5556
Fax (405) 329-2681
Oxine coating for sanitizing air ducts

Bio-Logical Controls Inc.
749 Hope Rd., Ste. A
Eatontown, NJ 07724
Tel (732) 389-8922; (800) 224-9768
Fax (732) 389-8821
Air purification systems

Boss Aire
2109 SE 4th Street
Minneapolis, MN 55414
Tel (800) 847-7390
Heat recovery ventilators, grilles, dampers

Broan Mfg. Co., Inc.
P.O. Box 140
Hartford, WI 53027
(800) 548-0790
Heat recovery ventilators, central and local ventilation equipment, controls

Carrier Corp.
P.O. Box 70
Indianapolis, IN 46206
Tel (800) 227-7437
Heat recovery ventilators, AC equipment, heat pumps

Clayton Environmental Consultants
41650 Gardenbrook Road, Suite 155
Novi, MI 48375
Tel (248) 344-8550
Fax (248) 344-0229
IAQ, safety, hazard evaluations

Clean-Aire Technologies
2327 Handley Ederville Road
P.O. Box 24638
Fort Worth, TX 76124
Tel (817) 589-7873
Fax (817) 595-0240
Duct cleaning and video inspection equipment

Columbus Industries
2938 State Road 752
Ashville, OH 43103
Tel (740) 983-2552
Fax (740) 983-4622
Polysorb activated carbon filters

Conservation Energy Systems, Inc.
2525 Wentz Avenue
Saskatoon, SK, Canada S7K 2K9
(306) 242-3663

Conservation Energy Systems, Inc.
Box 582416
Minneapolis, MN 55458
Tel (800) 667-3717
Heat recovery ventilators, controls, dampers, grilles, airflow-measuring equipment

Des Champs Laboratories, Inc.
Box 220
Natural Bridge Station, VA 24579
Tel (540) 291-1111
Heat recovery ventilators

Duct Sox
8900 N. Arbon Dr.
Milwaukee, WI 53223
Tel (800) 456-0600
Fax (414) 362-0661
Fabric ducts

Duro Dyne Corp.
130 Route 110
Farmingdale, NY 11735
Tel (800) 899-3876
Heat recovery ventilators, fresh-air intake control package

Dust Free, Inc.
P.O. Box 519
Royse City, TX 75189
Tel (800) 441-1107
Electrostatic air filters

Dwyer Instruments, Inc.
P.O. Box 373
Michigan City, IN 46360
Tel (219) 879-8000
Differential pressure transducers and air pressure gauges

Emerson Electric Co.
White Rodgers Division
1050 Dearborn
Columbus, OH 43229
Tel (800) 877-9222
Fax (641) 841-6022
Electrostatic precipitators

Energy Conservatory
5158 Bloomington Avenue
South Minneapolis, MN 55417
Tel (612) 827-1117
Blower doors, duct diagnostic fans, digital micromanometers

Environment Air Ltd.
P.O. Box 10
Cocagne, NB, Canada EOA I K0
Tel (506) 576-6672
Heat recovery ventilators, balanced ventilators, fans

Fabric Duct Systems Inc.
3530 E. 28th St.
Minneapolis, MN 55406
Tel (612) 721-2401
Fax (612) 721-0143
Permeable cloth ducts

Fan America, Inc.
1748 Independence Boulevard
Suite F-5
Sarasota, FL 34234
Tel (941) 359-3616
Fans and multiport ventilators

Fantech, Inc.
1712 Northgate Boulevard, Suite B
Sarasota, FL 34234
Tel (800) 747-1762
Central exhaust equipment, fans, wall vents, grilles, speed controls

Farr Co.
P.O. Box 92187
Los Angeles, CA 90009
(800) 333-7320
Air filters

First Alert
780 McClure Road
Aurora, IL 60504
Tel (800) 323-9005
Carbon monoxide, smoke detectors

G-Controls, Inc.
10734 Lake City Way NE
P.O. Box 27354
Seattle, WA 98125
Tel (206) 363-4863
Mixed-gas sensors

General Filters, Inc.
P.O. Box 8025
Novi, Ml 48376-8025
Tel (248) 476-5100
Fax (248) 349-2366
Medium-efficiency filters

Gollum International Inc.
305 Cayuga Road, Suite 140
Buffalo, N.Y. 14225
Tel (800) 546-5586
Duct cleaning equipment

Grainger, Inc.
333 Knightsbridge Parkway
Lincolnshire, IL 60069-3639
Fans, humidistats, timers

Grasslin Controls Corp.
24 Park Way
Upper Saddle River, NJ 07458
Tel (201) 825-9696
Timer controls

Greenheck
P.O. Box 410
Schofield, Wl 54476-0410
Tel (715) 359-6171
Ventilating fans

Hart & Cooley, Inc.
also Tuttle & Bailey
500 E. Eighth St.
Holland, MI 49423
Tel (616) 392-7855
Fax (800) 223-8461
Grilles, diffusers, registers, louvers, terminal units

Honeywell, Inc.
1985 Douglas Dr. North
Golden Valley, MN 55422-3992
Tel (800) 328-5111
Heat recovery ventilators, air cleaners, controls

HyCal
9650 Telstar Avenue
El Monte, CA 91731
Tel (818) 444-4000
Fax (818) 444-1314
CO_2 *sensors, controls, probes, transmitters*

ICC Technologies
441 N. 5th Street
Philadelphia, PA 19123
Tel (215) 625-0700
Fax (215) 592-8299
Desiccant dehumidification

Infiltec
P.O. Box 2956
Syracuse, NY 13220
Tel (800) 334-0837
Fax (315) 433-1521
Blower doors, duct diagnostic fans

Instrument Technology Inc.
P.O. Box 381
Westfield, MA 01086
413-562-3606; 413-568-9809
Boroscopes

Intermatic Inc.
Intermatic Plaza
Spring Grove, IL 60081-9698
Tel (815) 675-2321
Timer controls

Johnson Controls, Inc.
P.O. Box 423
Milwaukee, Wl 53201
Tel (414) 524-4000
Fax (414) 347-0221
Building automation, controls

Kanalflakt, Inc.
1712 Nonhgate Boulevard
Sarasota, FL 34234
Tel (941) 359-3267
Fax (941) 359-3828
Fans, grilles, controls, dampers

KD Engineering Inc.
No. 113-161 C. St.
Blaine, WA 98230
Tel (800) 308-7717
Fax (800) 739-4497
IAQ monitors

Kleen-Air Company, Inc.
269 West Caramel Drive
Carmel, IN 46032
Tel (317) 848-2757
Duct-mounted ozone generators

LakeAir International
1509 Rapids Drive
P.O. Box 4150
Racine, WI 53404
Tel (414) 632-1229
Fax (414) 632-3144
Electronic air cleaners and air filters

Landis & Gyr Powers Inc.
1000 Deerfield Parkway
Buffalo Grove, IL 60089
Tel (708) 215-1000
Fax (708) 215-1093
Temperature and energy management controls

Lennox Industries
2100 Lake Park Boulevard
Richardson, TX 75081
Tel (972) 497-5000
Heat recovery ventilators, AC equipment

Lindab Inc.
Two Stamford Landing
Stamford, CT 06902
Tel (203) 325-4666
Fax (203) 325-2111
SPIROsafe ducts

Meyer Machine and Equipment Inc.
241 Depot Street
Antioch, IL 60002
Tel (800) 728-3828
Duct cleaning equipment

Meyer & Sons Inc., Wm.
8261 Elmwood Avenue
P.O. Box 105
Tel (708) 673-0312
Fax (708) 673-5564
KleanKing duct cleaning equipment

Modus Instruments, Inc.
10 Bearfoot Road
Northboro, MA 01532
Tel (508) 393-8991
Pressure transducers and gauges

MSA Instrument
P.O. Box 427
Pittsburgh, PA 15230
Tel (800) MSA-2222
Fax (800) 967-0398
Refrigerant monitors, gas detectors

Munters Corp. Cargocaire Div.
79 Monroe St.
P.O. Box 640
Amesbury, MA 01913
Tel (508) 388-0600
Fax (508) 388-4556
Desiccant dehumidification

Newtron Products Co.
P.O. Box 27175
Cincinnati, OH 45227
Tel (800) 543-9149
Electrostatic air filters

Nutech Energy Systems, Inc.
511 McCormick Boulevard
London, Ont., Canada NSW 4C8
Tel (519) 457-1904
Heat recovery ventilators, air filters, grilles, controls

Penn Ventilator Co.
P.O. Box 52884
Philadelphia, PA 19115-7844
Tel (215) 464-8900
Fans, speed controls, time delay switches

Perfect Sense
992 S. Second St.
Ronkonkoma, NY 11779
Tel (631) 585-1005
Fax (631) 585-8643
Indoor air quality sensors

Permatron Corp.
11400 Melrose Street
Franklyn Park, IL 60131
Tel (800) 882-8012
Fax (847) 451-1811
Electrostatic air filters and carbon filters

Power Vac Inc.
639 South 29th Street
P.O. Box 341067
Milwaukee, WI 53234-1067
Tel (800) 822-4887
Duct cleaning equipment

Power Vacuum Trailer Company
1519 Cedar St.
Elmira, NY 14902
Tel (800) 382-8822
Fax (800) 711-4788
Duct cleaning equipment

Pringle, W.R.
P.O. Box 306
Napoleon, OH 43545
Tel (419) 256-7763
Fax (419) 592-3402
Duct cleaning equipment

Purafil, Inc.
P.O. Box 1188
Norcross, GA 30091
Tel (404) 662-8545
Adsorption material

Purolator Products
P.O. Box 1637
Henderson, N.C. 27536
Tel (252) 494-1141
Fax (252) 492-6157
Air filters

Raydot, Inc.
145 Jackson Avenue
Cokato, MN 55321
Tel (800) 328-3813
Fax (608) 257-4357
Heat recovery ventilators, filtration module

Research Products Corp.
P.O. Box 1467
Madison, WI 53701-1467
Tel (800) 334-6011
Fax 608-257-4357
Heat recovery ventilators, air cleaners, humidifiers

Retrotec, Inc.
12-2200 Queen Street
Bellingham, WA 98226-4766
Tel (206) 738-9835
Blower doors

RSE, Inc.
P.O. Box 26
New Baltimore, MI 48047-0026
Tel (810) 725-0192
Activated carbon filters

Ruskin Co.
3900 Dr. Greaves Rd.
Kansas City, MO 64030
Tel (816) 761-7476
Fax (816) 761-0521
Dampers, air monitors

SEMCO Inc.
1800 East Pointe Drive
Columbia, MO 65201
Tel (573) 443-1481
Fax (573) 443-6921
Energy recovery, desiccant systems

Shortridge Instruments, Inc.
7855 E. Redfield Road
Scottsdale, AZ 85260
Tel (408) 991-6744
Airflow measurement devices

Skuttle IAQ Products
Route 10
Marietta, OH 45750-9990
Tel (740) 373-9169
Make-up air intakes

Solomat Neotronics
26 Pearl, Waterside Bldg.
Norwalk, CT 06850
Tel (203) 849-3111
Air quality measuring instruments

Specialized Ventilation Systems
Rt. 3 Box 388M
Buffalo, MO 65622-9803
Tel (417) 345-5329
Fax (417) 345-5435
Antimicrobial ultraviolet lighting

Staefa Control System, Inc.
1000 Deerfield Pkwy.
Buffalo Grove, IL 60089-4547
Tel (847) 215-1050; (800) 735-4822
Fax (847) 941-4814
Mixed-gas sensors

Steril-Aire USA Inc.
11100 E. Arresia Blvd., #D
Cerritos, CA 90703
Tel (562) 467-8484
Fax (562-467-8481
Antimicrobial ultraviolet lighting

Stirling Technology Inc.
P.O. Box 2633
Athens, OH 45701
Tel (614) 592-2277
Fax (614) 592-1499
Heat recovery ventilators

Stulz Air Technology Systems Inc.
1572 Tilco Dr.
Frederick, MD 21701
Tel (301) 620-2033
Fax (301) 662-5487
Ultrasonic humidification

Telaire Systems, Inc.
6489 Calle Real
Goleta, CA 93117
Tel (805) 964-1699
Fax (805) 964-2129
Carbon dioxide controller, diagnostic instruments

Thermax
P.O. Box 300
Hopewell, NJ 08525-0300
Tel (800) 929-0682
Heat recovery ventilators, central and window units

Therma-Stor Products
P.O. Box 8050
2001 S. Stoughton Rd.
Madison, WI 53708
Tel (800) 533-7533
*Air filters, heat-recovery
        water heaters, dehumidifiers*

Titon Inc.
P.O.Box 6164
South Bend, IN 46660
Tel (219) 271-9699
Through-the-wall vents

Tjernlund Products, Inc.
1601 9th Street
White Bear Lake, MN 55110-6794
Tel (800) 255-4208
Fresh-air intake systems

Tork
1 Grove Street
Mount Vernon, NY 10550
Tel (914) 664-3542
Fax (914) 664-5052
Controls

Trion, Inc.
P.O. Box 760
Sanford, NC 27331-0760
Tel (919) 775-2201
Fax (919) 777-6399
Electrostatic precipitators,
electronic air cleaners

TSI Inc.
500 Cardigan Road
P.O. Box 64394
Saint Paul, MN 55164
Tel (800) 876-9874
IAQ monitors and instruments

Vac System Industries
5995 195th Street West, Suite 102
Apple Valley, MN 55124-5711
Tel (952) 432-3955
Fax (952) 432-1584
Duct cleaning equipment

Venmar CES, Inc.
2525 Wentz Ave.
Saskatoon, SK S7K 2K9
Canada
Tel (800) 667-3717
Fax (306) 242-3484
Heat recovery

Warren Technology
2050 W. 73rd Street
P.O. Box 5347
Hialeah, FL 33016
Tel (305) 556-6933
Fax (305) 557-6157
Diffusers, dampers, air handlers

WaterFurnace International
9000 Conservation Way
Fort Wayne, IN 46809
Tel (260) 478-5667
Fax (260) 478-3029
Geothermal heat pumps

York International
P.O. Box 1592-232-A
York, PA 17405
Tel (405) 364-4040
Fax (405) 419-6545
Chillers, air handlers and other HVAC equipment

Appendix IV

IAQ Forms for Practitioners and Facilities Managers

Forms are courtesy of the
U.S. Environmental Protection Agency.

IAQ Management Checklist

Building Name: _____ Date: _____

Address: _____

Completed by (name/title): _____

Use this checklist to make sure that you have included all necessary elements in your IAQ profile and IAQ management plan. *Sections 4 and 5* discuss the development of the IAQ profile and IAQ management plan.

Item	Date begun or completed (as applicable)	Responsible person (name, telephone)	Location ("NA" if the item is not applicable to this building)
IAQ PROFILE			
Collect and Review Existing Records			
HVAC design data, operating instructions, and manuals			
HVAC maintenance and calibration records, testing and balancing reports			
Inventory of locations where occupancy, equipment, or building use has changed			
Inventory of complaint locations			
Conduct a Walkthrough Inspection of the Building			
List of responsible staff and/or contractors, evidence of training, and job descriptions			
Identification of areas where positive or negative pressure should be maintained			
Record of locations that need monitoring or correction			
Collect Detailed Information			
Inventory of HVAC system components needing repair, adjustment, or replacement			
Record of control settings and operating schedules			

IAQ Management Checklist

Item	Date begun or completed (as applicable)	Responsible person (name, telephone)	Location ("NA" if the item is not applicable to this building)
Plan showing airflow directions or pressure differentials in significant areas			
Inventory of significant pollutant sources and their locations			
MSDSs for supplies and hazardous substances that are stored or used in the building			
Zone/Room Record			
IAQ MANAGEMENT PLAN			
Select IAQ Manager			
Review IAQ Profile			
Assign Staff Responsibilities Track Staff			
Facilities Operation and Maintenance			
■ confirm that equipment operating schedules are appropriate			
■ confirm appropriate pressure relationships between building usage areas			
■ compare ventilation quantities to design, codes, and ASHRAE 62-1989			
■ schedule equipment inspections per preventive maintenance plan or recommended maintenance schedule			
■ modify and use HVAC Checklist(s); update as equipment is added, removed, or replaced			
■ schedule maintenance activities to avoid creating IAQ problems			

IAQ Management Checklist

Item	Date begun or completed (as applicable)	Responsible person (name, telephone)	Location ("NA" if the item is not applicable to this building)
■ review MSDSs for supplies; request additional information as needed			
■ consider using alarms or other devices to signal need for HVAC maintenance (e.g., clogged filters)			
Housekeeping			
■ evaluate cleaning schedules and procedures; modify if necessary			
■ review MSDSs for products in use; buy different products if necessary			
■ confirm proper use and storage of materials			
■ review trash disposal procedures; modify if necessary			
Shipping and Receiving			
■ review loading dock procedures (*Note:* If air intake is located nearby, take precautions to prevent intake of exhaust fumes.)			
■ check pressure relationships around loading dock			
Pest Control			
■ consider adopting IPM methods			
■ obtain and review MSDSs; review handling and storage			
■ review pest control schedules and procedures			
■ review ventilation used during pesticide application			

IAQ Management Checklist

Item	Date begun or completed (as applicable)	Responsible person (name, telephone)	Location ("NA" if the item is not applicable to this building)
Occupant Relations			
■ establish health and safety committee or joint tenant/ management IAQ task force			
■ review procedures for responding to complaints; modify if necessary			
■ review lease provisions; modify if necessary			
Renovation, Redecorating, Remodeling			
■ discuss IAQ concerns with architects, engineers, contractors, and other professionals			
■ obtain MSDSs; use materials and procedures that minimize IAQ problems			
■ schedule work to minimize IAQ problems			
■ arrange ventilation to isolate work areas			
■ use installation procedures that minimize emissions from new furnishings			
Smoking			
■ eliminate smoking in the building			
■ if smoking areas are designated, provide adequate ventilation and maintain under negative pressure			
■ work with occupants to develop appropriate non-smoking policies, including implementation of smoking cessation programs			

Pollutant Pathway Record For
IAQ Profiles

This form should be used in combination with a floor plan such as a fire evacuation plan.

Building Name: _____ File Number: _____

Address: _____

Completed by: _____ Title: _____ Date: _____

Sections 2, 4 and 6 discuss pollutant pathways and driving forces.

Building areas that contain contaminant sources (e.g., bathrooms, food preparation areas, smoking lounges, print rooms, and art rooms) should be maintained under negative pressure relative to surrounding areas. Building areas that need to be protected from the infiltration of contaminants (e.g., hallways in multi-family dwellings, computer rooms, and lobbies) should be maintained under positive pressure relative to the outdoors and relative to surrounding areas.

List the building areas in which pressure relationships should be controlled. As you inspect the building, put a Y or N in the "Needs Attention" column to show whether the desired air pressure relationship is present. Mark the floor plan with arrows, plus signs (+) and minus signs (-) to show the airflow patterns you observe using chemical smoke or a micromanometer.

Building areas that appear isolated from each other may be connected by airflow passages such as air distribution zones, utility tunnels or chases, party walls, spaces above suspended ceilings (whether or not those spaces are serving as air plenums), elevator shafts, and crawlspaces. If you are aware of pathways connecting the room to identified pollutant sources (e.g., items of equipment, chemical storage areas, bathrooms), it may be helpful to record them in the "Comments" column, on the floor plan, or both.

Building Area (zone, room)	Use	Intended Pressure		Needs Attention? (Y/N)	Comments
		Positive (+)	Negative (-)		

Zone/Room Record

Building Name: _____ File Number: _____ Date: _____

Address: _____ Completed by: _____ Title: _____

This form is to be used differently depending on whether the goal is to *prevent* or to *diagnose* IAQ problems. During the development of a profile, this form should be used to record more general information about the entire building; during an investigation, the form should be used to record more detailed information about the complaint area and areas surrounding the complaint area or connected to it by pathways.

Use the last three columns when underventilation is suspected. Use the **Ventilation Worksheet** and *Appendix A* to estimate outdoor air quantities. Compare results to the design specifications, applicable building codes, or ventilation guidelines such as ASHRAE 62-1989. (See *Appendix A* for some outdoor air quantities required by ASHRAE 62-1989.) *Note:* For VAV systems, minimum outdoor air under reduced flow conditions must be considered.

Building Area (Zone/Room)	Use**	PROFILE AND DIAGNOSIS INFORMATION			DIAGNOSIS INFORMATION ONLY		
		Source of Outdoor Air*	Mechanical Exhaust? (Write "No" or estimate cfm airflow)	Comments	Peak Number of Occupants or Sq. Ft. Floor Area**	Total Air Supplied (in cfm)***	Outdoor Air Supplied per Person or per 150 Sq. Ft. Area (in cfm)****

* Sources might include air handling unit (e.g., AHU-4), operable windows, transfer from corridors
** Underline the information in this column if current use or number of occupants is different from design specifications
*** Mark the information with a **P** if it comes from the mechanical plans or an **M** if it comes from the actual measurements, such as recent test and balance reports.
**** ASHRAE 62-1989 gives ventilation guidance per 150 sq. ft.

Ventilation Worksheet

Building Name: _____ File Number: _____

Address: _____

Completed by (name): _____ Date: _____

This worksheet is designed for use with the **Zone/Room Record.** *Appendix A* provides guidance on methods of estimating the amount of ventilation (outdoor) air being introduced by a particular air handling unit. *Appendix B* discusses the ventilation recommendations of ASHRAE Standard 62-1989, which was developed for the purpose of preventing indoor air quality problems. Formulas are given below for calculating outdoor air quantities using thermal or CO_2 information.

The equation for calculating outdoor air quantities **using thermal measurements** is:

$$\text{Outdoor air (in percent)} = \frac{T_{\text{return air}} - T_{\text{mixed air}}}{T_{\text{return air}} - T_{\text{outdoor air}}} \times 100$$

Where: T = temperature in degrees Fahrenheit

The equation for calculating outdoor quantities **using carbon dioxide measurements** is:

$$\text{Outdoor air (in percent)} = \frac{C_s - C_r}{C_0 - C_r} \times 100$$

Where: C_s = ppm of carbon dioxide in the supply air (if measured in a room), or
C_s = ppm of carbon dioxide in the mixed air (if measured at an air handler)
C_r = ppm of carbon dioxide in the return air
C_o = ppm of carbon dioxide in the outdoor air

Use the table below to estimate the ventilation rate in any room or zone. *Note:* ASHRAE 62-1989 generally states ventilation (outdoor air) requirements on an occupancy basis; for a few types of spaces, however, requirements are given on a floor area basis. Therefore, this table provides a process of calculating ventilation (outdoor air) on either an occupancy or floor area basis.

Zone/Room	Percent of Outdoor Air	Total Air Supplied to Zone/Room (cfm)	Peak Occupancy (number of people) or Floor Area (square feet)	$D = \frac{B}{C}$ Total Air Supplied Per Person (or per square foot area)	$E = (A \times 100) \times D$ Outdoor air Supplied Per Person (or per square foot area)
	A	B	C	D	E

Indoor Air Quality Complaint Form

This form can be filled out by the building occupant or by a member of the building staff.

Occupant Name: _____ Date: _____

Department/Location in Building: _____ Phone: _____

Completed by: _____ Title: _____ Phone: _____

This form should be used if your complaint may be related to indoor air quality. Indoor air quality problems include concerns with temperature control, ventilation, and air pollutants. Your observations can help to resolve the problem as quickly as possible. Please use the space below to describe the nature of the complaint and any potential causes.

We may need to contact you to discuss your complaint. What is the best time to reach you? _____

So that we can respond promptly, please return this form to: _____
                                                            IAQ Manager or Contact Person

                                                            _____
                                                            Room, Building, Mail Code

OFFICE USE ONLY

File Number: _____ Received By: _____ Date Received: _____

Incident Log

Building Name: _____   Dates (from): _____   (to): _____

Address: _____   Completed by (name): _____

| File Number | Date | Problem Location | Investigation Record (check the forms that were used) | | | | | | | | | Outcome/Comments (use more than one line if needed) | Log Entry By (initials) |
			Complaint Form	Occupant Interview	Occupant Diary	Log of Activities	Zone/Room Record	HVAC Checklist	Pollutant Pathway	Source Inventory	Hypothesis Form		

Occupant Interview

Building Name: _____ File Number: _____

Address: _____

Occupant Name: _____ Work Location: _____

Completed by: _____ Title: _____ Date: _____

Section 4 discusses collecting and interpreting information from occupants.

SYMPTOM PATTERNS

What kind of symptoms or discomfort are you experiencing?

Are you aware of other people with similar symptoms or concerns? Yes _____ No _____

If so, what are their names and locations? _____

_____

Do you have any health conditions that may make you particularly susceptible to environmental problems?

❑ contact lenses   ❑ chronic cardiovascular disease   ❑ undergoing chemotherapy or radiation therapy

❑ allergies   ❑ chronic respiratory disease   ❑ immune system suppressed by disease or
                                                              other causes

                      ❑ chronic neurological problems

TIMING PATTERNS

When did your symptoms start?

When are they generally worst?

Do they go away? If so, when?

Have you noticed any other events (such as weather events, temperature or humidity changes, or activities in the building) that tend to occur around the same time as your symptoms?

Occupant Interview

SPATIAL PATTERNS

Where are you when you experience symptoms or discomfort?

Where do you spend most of your time in the building?

ADDITIONAL INFORMATION

Do you have any observations about building conditions that might need attention or might help explain your symptoms (e.g., temperature, humidity, drafts, stagnant air, odors)?

Have you sought medical attention for your symptoms?

Do you have any other comments?

Occupant Diary

Occupant Name: _____ Title: _____ Phone: _____

Location: _____ File Number :_____

On the form below, please record each occasion when you experience a symptom of ill-health or discomfort that you think may be linked to an environmental condition in this building.

It is important that you record the time and date and your location within the building as accurately as possible, because that will help to identify conditions (e.g., equipment operation) that may be associated with your problem. Also, please try to describe the severity of your symptoms (e.g., mild, severe) and their duration (the length of time that they persist). Any other observations that you think may help in identifying the cause of the problem should be noted in the "Comments" column. Feel free to attach additional pages or use more than one line for each event if you need more room to record your observations.

Section 6 discusses collecting and interpreting occupant information.

Time/Date	Location	Symptom	Severity/Duration	Comments

Log of Activities and System Operation

Building Name: _____ Address: _____ File Number : _____

Completed by: _____ Title: _____ Phone: _____

On the form below, please record your observations of the HVAC system operation, maintenance activities, and any other information that you think might be helpful in identifying the cause of IAQ complaints in this building. Please report any other observations (e.g., weather, other associated events) that you think may be important as well.

Feel free to attach additional pages or use more than one line for each event.

Equipment and activities of particular interest:
Air Handler(s): _____
Exhaust Fan(s): _____
Other Equipment or Activities: _____

Date/Time	Day of Week	Equipment Item/Activity	Observations/Comments

HVAC Checklist - Short Form

Building Name: _____ Address: _____

Completed by: _____ Date: _____ File Number: _____

Sections 2, 4 and 6 and Appendix B discuss the relationships between the HVAC system and indoor air quality.

MECHANICAL ROOM

- Clean and dry? _____ Stored refuse or chemicals? _____
- Describe items in need of attention _____

MAJOR MECHANICAL EQUIPMENT

- Preventive maintenance (PM) plan in use? _____

Control System

- Type _____
- System operation _____
- Date of last calibration _____

Boiler

- Rated Btu input _____ Condition _____
- Combustion air: is there at least one square inch free area per 2,000 Btu input? _____
- Fuel or combustion odors _____

Cooling Tower

- Clean? no leaks or overflow? _____ Slime or algae growth? _____
- Eliminator performance _____
- Biocide treatment working? (list type of biocide) _____
- Spill containment plan implemented? _____ Dirt separator working? _____

Chillers

- Refrigerant leaks? _____
- Evidence of condensation problems? _____
- Waste oil and refrigerant properly stored and disposed of? _____

HVAC Checklist - Short Form

Building Name: _____ Address: _____

Completed by: _____ Date: _____ File Number: _____

AIR HANDLING UNIT

■ Unit identification _____ Area served _____

Outdoor Air Intake, Mixing Plenum, and Dampers

■ Outdoor air intake location _____

■ Nearby contaminant sources? (describe) _____

■ Bird screen in place and unobstructed? _____

■ Design total cfm _____ outdoor air (O.A.) cfm _____ date last tested and balanced _____

■ Minimum % O.A. (damper setting) _____ Minimum cfm O.A. $\dfrac{\text{(total cfm} \times \text{minimum \% O.A.)}}{100}$ = _____

■ Current O.A. damper setting (date, time, and HVAC operating mode) _____

■ Damper control sequence (describe) _____

■ Condition of dampers and controls (note date) _____

Fans

■ Control sequence _____

■ Condition (note date) _____

■ Indicated temperatures   supply air _____ mixed air _____ return air _____ outdoor air _____

■ Actual temperatures   supply air _____ mixed air _____ return air _____ outdoor air _____

Coils

■ Heating fluid discharge temperature _____ ΔT _____ cooling fluid discharge temperature _____ ΔT _____

■ Controls (describe) _____

■ Condition (note date) _____

Humidifier

■ Type _____ If biocide is used, note type _____

■ Condition (no overflow, drains trapped, all nozzles working?) _____

■ No slime, visible growth, or mineral deposits? _____

HVAC Checklist - Short Form

Building Name: _____ Address: _____

Completed by: _____ Date: _____ File Number: _____

DISTRIBUTION SYSTEM

Zone/ Room	System Type	Supply Air		Return Air		Power Exhaust		
		ducted/ unducted	cfm	ducted/ unducted	cfm	cfm	control	serves (e.g. toilet)

Condition of distribution system and terminal equipment (note locations of problems)

- Adequate access for maintenance? _____
- Ducts and coils clean and obstructed? _____
- Air paths unobstructed?   supply _____ return _____ transfer _____ exhaust _____ make-up _____
- Note locations of blocked air paths, diffusers, or grilles _____
- Any unintentional openings into plenums? _____
- Controls operating properly? _____
- Air volume correct? _____
- Drain pans clean? Any visible growth or odors? _____

Filters

Location	Type/Rating	Size	Date Last Changed	Condition (give date)

HVAC Checklist - Short Form

Building Name: _____ Address: _____

Completed by: _____ Date: _____ File Number: _____

OCCUPIED SPACE

Thermostat types

Zone/ Room	Thermostat Location	What Does Thermostat Control? (e.g., radiator, AHU-3)	Setpoints		Measured Temperature	Day/ Time
			Summer	Winter		

Humidistat/Dehumidistat types

Zone/ Room	Humidistat/ Dehumidistat Location	What Does It Control?	Setpoints (%RH)	Measured Temperature	Day/ Time

■ Potential problems (note location) _____

■ Thermal comfort or air circulation problems (drafts, obstructed airflow, stagnant air, overcrowding, poor thermostat location)

_____

_____

■ Malfunctioning equipment _____

■ Major sources of odors or contaminants (e.g., poor sanitation, incompatible uses of space)

_____

HVAC Checklist - Long Form

Building: _____ File Number: _____

Completed by: _____ Title: _____ Date Checked: _____

Appendix B discusses HVAC system components in relation to indoor air quality.

Component	OK	Needs Attention	Not Applicable	Comments
Outside Air Intake				
Location _____				
Open during occupied hours?				
Unobstructed?				
Standing water, bird droppings in vicinity?				
Odors from outdoors? (describe) _____				
Carryover of exhaust heat?				
Cooling tower within 25 feet?				
Exhaust outlet within 25 feet?				
Trash compactor within 25 feet?				
Near parking facility, busy road, loading dock?				
Bird Screen				
Unobstructed?				
General condition?				
Size of mesh? ($1/2$" minimum)				
Outside Air Dampers				
Operation acceptable?				
Seal when closed?				

HVAC Checklist - Long Form

Building: _____ File Number: _____

Completed by: _____ Title: _____ Date Checked: _____

Component	OK	Needs Attention	Not Applicable	Comments
Actuators operational?				
Outdoor Air (O.A.) Quantity (Check against applicable codes and ASHRAE 62-1989.)				
Minimum % O.A. ————————				
Measured % O.A. ———————— _Note day, time, HVAC operating mode under "Comments"_				
Maximum % O.A. ————————				
Is minimum O.A. a separate damper?				
For VAV systems: is O.A. increased as total system air-flow is reduced?				
Mixing Plenum				
Clean?				
Floor drain trapped?				
Airtightness				
▪ of outside air dampers				
▪ of return air dampers				
▪ of exhaust air dampers				
All damper motors connected?				
All damper motors operational?				
Air mixers or opposed blades?				

HVAC Checklist - Long Form

Building: _____ File Number: _____

Completed by: _____ Title: _____ Date Checked: _____

Component	OK	Needs Attention	Not Applicable	Comments
Mixed air temperature control setting _____°F				
Freeze stat setting _____°F				
Is mixing plenum under negative pressure? *Note: If it is under positive pressure, outdoor air may not be entering.*				
Filters				
Type _____				
Complete coverage? (i.e., no bypassing)				
Correct pressure drop? *(Compare to manufacturer's recommendations.)*				
Contaminants visible?				
Odor noticeable?				
Spray Humidifiers or Air Washers				
Humidifier type				
All nozzles working?				
Complete coil coverage?				
Pans clean, no overflow?				
Drains trapped?				
Biocide treatment working? *Note: Is MSDS on file?*_____				
Spill contaminant system in place?				

HVAC Checklist - Long Form

Building: _____   File Number: _____

Completed by: _____   Title: _____   Date Checked: _____

Component	OK	Needs Attention	Not Applicable	Comments
Face and Bypass Dampers				
Damper operation correct?				
Damper motors operational?				
Cooling Coils				
Inspection access?				
Clean?				
Supply water temp. ____°F				
Water carryover?				
Any indication of condensation problems?				
Condensate Drip Pans				
Accessible to inspect and clean?				
Clean, no residue?				
No standing water, no leaks?				
Noticeable odor?				
Visible growth (e.g., slime)?				
Drains and traps clear, working?				
Trapped to air gap?				
Water overflow?				

HVAC Checklist - Long Form

Building: _____ File Number: _____

Completed by: _____ Title: _____ Date Checked: _____

Component	OK	Needs Attention	Not Applicable	Comments
Mist Eliminators				
Clean, straight, no carryover?				
Supply Fan Chambers				
Clean?				
No trash or storage?				
Floor drain traps are wet or sealed?				
No air leaks?				
Doors close tightly?				
Supply Fans				
Location _____				
Fan blades clean?				
Belt guards installed?				
Proper belt tension?				
Excess vibration?				
Corrosion problems?				
Controls operational, calibrated?				

HVAC Checklist - Long Form

Building: _____ File Number: _____

Completed by: _____ Title: _____ Date Checked: _____

Component	OK	Needs Attention	Not Applicable	Comments
Control sequence conforms to design/specifications? (describe changes)				
No pneumatic leaks?				
Heating Coil				
Inspection access?				
Clean?				
Control sequence conforms to design/specifications? (describe changes)				
Supply water temp. _____°F				
Discharge thermostat? (air temp. setting _____°F)				
Reheat Coils				
Clean?				
Obstructed?				
Operational?				
Steam Humidifier				
Humidifier type _____				
Treated boiler water?				
Standing water?				

HVAC Checklist - Long Form

Building: _____   File Number: _____

Completed by: _____ Title: _____ Date Checked: _____

Component	OK	Needs Attention	Not Applicable	Comments
Visible growth?				
Mineral deposits?				
Control setpoint _____°F				
High limit setpoint _____°F				
Duct liner within 12 feet? (If so, check for dirt, mold growth.)				
Supply Ductwork				
Clean?				
Sealed, no leaks, tight connections?				
Fire dampers open?				
Access doors closed?				
Lined ducts?				
Flex duct connected, no tears?				
Light troffer supply?				
Balanced within 3-5 years?				
Balanced after recent renovations?				
Short circuiting or other air distribution problems? Note location(s) _____				
Pressurized Calling Supply Plenum				
No unintentional openings?				
All ceiling tiles in place?				

HVAC Checklist - Long Form

Building: _____   File Number: _____

Completed by: _____   Title: _____   Date Checked: _____

Component	OK	Needs Attention	Not Applicable	Comments
Barrier paper correctly placed and in good condition?				
Proper layout for air distribution?				
Supply diffusers open?				
Supply diffusers balanced?				
Balancing capability?				
Noticeable flow of air?				
Short circuiting or other air distribution problems? *Note location(s) in"Comments"*				
Terminal Equipment (supply)				
Housing interiors clean and unobstructed?				
Controls working?				
Delivering rated volume?				
Balanced within 3-5 years?				
Filters in place?				
Condensate pans clean, drain freely?				
VAV Box				
Minimum stops _____ %				
Minimum outside air ____ % *(from page 2 of this form)*				
Minimum airflow _____cfm				
Minimum outside air_____ cfm				

HVAC Checklist - Long Form

Building: _____ File Number: _____

Completed by: _____ Title: _____ Date Checked: _____

Component	OK	Needs Attention	Not Applicable	Comments
Supply setpoint _____ °F (summer) _____ °F (winter)				
Thermostats				
Type _____				
Properly located?				
Working?				
Setpoints _____ °F (summer) _____ °F (winter)				
Space temperature _____ °F				
Humidity Sensor				
Humidistat setpoints _____ % RH				
Dehumidistat setpoints _____ % RH				
Actual RH _____ %				
Room Partitions				
Gap allowing airflow at top?				
Gap allowing airflow at bottom?				
Supply and return each room?				

HVAC Checklist - Long Form

Building: _____ File Number: _____

Completed by: _____ Title: _____ Date Checked: _____

Component	OK	Needs Attention	Not Applicable	Comments
Stairwells				
Doors close and latch?				
No openings allowing uncontrolled airflow?				
Clean, dry?				
No noticeable odors?				
Return Air Plenum				
Tiles in place?				
No unintentional openings?				
Return grilles?				
Balancing capability?				
Noticeable flow of air?				
Transfer grilles?				
Fire dampers open?				
Ducted Returns				
Balanced within 3-5 years?				
Unobstructed grilles?				
Unobstructed return air path?				
Return Fan Chambers				
Clean and no trash or storage?				
No standing water?				
Floor drain traps are wet or sealed?				

HVAC Checklist - Long Form

Building: _____ File Number: _____

Completed by: _____ Title: _____ Date Checked: _____

Component	OK	Needs Attention	Not Applicable	Comments
No air leaks?				
Doors close tightly, kept closed?				
Return Fans				
Location _____				
Fan blades clean?				
Belt guards installed?				
Proper belt tension?				
Excess vibration?				
Corrosion problems?				
Controls working, calibrated?				
Control sequence conforms to design/specifications? (describe changes)				
Exhaust Fans				
Central?				
Distributed (locations) _____ _____				
Operational?				
Controls operational?				
Toilet exhaust only?				
Gravity relief?				

HVAC Checklist - Long Form

Building: _____   File Number: _____

Completed by: _____   Title: _____   Date Checked: _____

Component	OK	Needs Attention	Not Applicable	Comments
Total powered exhaust _____ cfm				
Make-up air sufficient?				
Toilet Exhausts				
Fans working occupied hours?				
Registers open, clear?				
Make-up air path adequate?				
Volume according to code?				
Floor drain traps wet or sealable?				
Bathrooms run slightly negative relative to building?				
Smoking Lounge Exhaust				
Room runs negative relative to building?				
Print Room Exhaust				
Room runs negative relative to building?				
Garage Ventilation				
Operates according to codes?				
Fans, controls, dampers all operate?				

HVAC Checklist - Long Form

Building: _____ File Number: _____

Completed by: _____ Title: _____ Date Checked: _____

Component	OK	Needs Attention	Not Applicable	Comments
Garage slightly negative relative to building?				
Doors to building close tightly?				
Vestibule entrance to building from garage?				
Mechanical Rooms				
General condition?				
Controls operational?				
Pneumatic controls:				
▪ compressor operational?				
▪ air dryer operational?				
Electric controls? Operational?				
EMS (Energy Management System) or DDC (Direct Digital Control):				
▪ operator on site?				
▪ controlled off-site?				
▪ are fans cycled "off" while building is occupied?				
▪ is chiller reset to shed load?				
Preventive Maintenance				
Spare parts inventoried?				
Spare air filters?				
Control drawing posted?				

HVAC Checklist - Long Form

Building: _____   File Number: _____

Completed by: _____   Title: _____   Date Checked: _____

Component	OK	Needs Attention	Not Applicable	Comments
PM (Preventive Maintenance) schedule available?				
PM followed?				
Boilers				
Flues, breeching tight?				
Purge cycle working?				
Door gaskets tight?				
Fuel system tight, no leaks?				
Combustion air: at least 1 square inch free area per 2000 Btu input?				
Cooling Tower				
Sump clean?				
No leaks, no overflow?				
Eliminators working, no carryover?				
No slime or algae?				
Biocide treatment working?				
Dirt separator working?				
Chillers				
No refrigerant leaks?				
Purge cycle normal?				
Waste oil, refrigerant properly disposed of and spare refrigerant properly stored?				
Condensation problems?				

HVAC Checklist - Long Form

Building: _____ File Number: _____

Completed by: _____ Title: _____ Date Checked: _____

Component	OK	Needs Attention	Not Applicable	Comments

Pollutant Pathway Form For Investigations

Building Name: _____   File Number: _____

Address: _____   Completed by: _____

This form should be used in combination with a floor plan such as a fire evacuation plan.

Building areas that appear isolated from each other may be connected by airflow passages such as air distribution zones, utility tunnels or chases, party walls, spaces above suspended ceilings (whether or not those spaces are serving as air plenums), elevator shafts, and crawl spaces.

Describe the complaint area in the space below and mark it on your floor plan. Then list rooms or zones connected to the complaint area by airflow pathways. Use the form to record the direction of air flow between the complaint area and the connected rooms/zones, including the date and time. (Airflow patterns generally change over time). Mark the floor plan with arrows or plus (+) and minus (-) signs to map out the airflow patterns you observe, using chemical smoke or a micromanometer. The "Comments" column can be used to note pollutant sources that merit further attention.

Rooms or zones included in the complaint area: _____

_____

Sections 2, 4 and 6 discuss pollutant pathways and driving forces.

Rooms or Zones Connected to the Complaint Area By Pathways	Use	Pressure Relative to Complaint Area		Comments (e.g., potential pollutant sources)
		+/-	date/time	

Pollutant and Source Inventory

Building Name: _____ Address: _____

Completed by: _____ Date: _____ File Number: _____

Using the list of potential source categories below, record any indications of contamination or suspected pollutants that may require further investigation or treatment. Sources of contamination may be constant or intermittent or may be linked to single, unrepeated events. For intermittent sources, try to indicate the time of peak activity or contaminant production, including correlations with weather (e.g., wind direction).

Sections 2, 4 and 6 discuss pollutant sources. Appendix A provides guidance on common measurements.

Source Category	Checked	Needs Attention	Location	Comments
SOURCES OUTSIDE BUILDING				
Contaminated Outdoor Air				
Pollen, dust				
Industrial contaminants				
General vehicular contaminants				
Emissions from Nearby Sources				
Vehicle exhaust (parking areas, loading docks, roads)				
Dumpsters				
Re-entrained exhaust				
Debris near outside air intake				
Soil Gas				
Radon				
Leaking underground tanks				
Sewage smells				
Pesticides				

Pollutant and Source Inventory

Building Name: _____ Address: _____

Completed by: _____ Date: _____ File Number: _____

Using the list of potential source categories below, record any indications of contamination or suspected pollutants that may require further investigation or treatment. Sources of contamination may be constant or intermittent or may be linked to single, unrepeated events. For intermittent sources, try to indicate the time of peak activity or contaminant production, including correlations with weather (e.g., wind direction).

Source Category	Checked	Needs Attention	Location	Comments
Moisture or Standing Water				
Rooftop				
Crawlspace				
EQUIPMENT				
HVAC System Equipment				
Combustion gases				
Dust, dirt, or microbial growth in ducts				
Microbial growth in drip pans, chillers, humidifiers				
Leaks of treated boiler water				
Non HVAC System Equipment				
Office Equipment				
Supplies for Equipment				
Laboratory Equipment				

Pollutant and Source Inventory

Page 3 of 6

Building Name: _____ Address: _____

Completed by: _____ Date: _____ File Number: _____

Using the list of potential source categories below, record any indications of contamination or suspected pollutants that may require further investigation or treatment. Sources of contamination may be constant or intermittent or may be linked to single, unrepeated events. For intermittent sources, try to indicate the time of peak activity or contaminant production, including correlations with weather (e.g., wind direction).

Source Category	Checked	Needs Attention	Location	Comments
HUMAN ACTIVITIES				
Personal Activities				
Smoking				
Cosmetics (odors)				
Housekeeping Activities				
Cleaning materials				
Cleaning procedures (e.g., dust from sweeping, vacuuming)				
Stored supplies				
Stored refuse				
Maintenance Activities				
Use of materials with volatile compounds (e.g., paint, caulk, adhesives)				
Stored supplies with volatile compounds				
Use of pesticides				

Pollutant and Source Inventory

Building Name: _____ Address: _____

Completed by: _____ Date: _____ File Number: _____

Using the list of potential source categories below, record any indications of contamination or suspected pollutants that may require further investigation or treatment. Sources of contamination may be constant or intermittent or may be linked to single, unrepeated events. For intermittent sources, try to indicate the time of peak activity or contaminant production, including correlations with weather (e.g., wind direction).

Source Category	Checked	Needs Attention	Location	Comments
BUILDING COMPONENTS FURNISHINGS				
Locations Associated with Dust or Fibers				
Dust-catching area (e.g., open shelving)				
Deteriorated furnishings				
Asbestos-containing materials				
Unsanitary Conditions/Water Damage				
Microbial growth in or on soiled or water-damaged furnishings				

Pollutant and Source Inventory

Page 5 of 6

Building Name: _____ Address: _____

Completed by: _____ Date: _____ File Number: _____

Using the list of potential source categories below, record any indications of contamination or suspected pollutants that may require further investigation or treatment. Sources of contamination may be constant or intermittent or may be linked to single, unrepeated events. For intermittent sources, try to indicate the time of peak activity or contaminant production, including correlations with weather (e.g., wind direction).

Source Category	Checked	Needs Attention	Location	Comments
Chemicals Released from Building Components or Furnishings				
Volatile compounds				
OTHER SOURCES				
Accidental Events				
Spills (e.g., water, chemicals, beverages)				
Water leaks or flooding				
Fire damage				

Pollutant and Source Inventory

Building Name: _____ Address: _____

Completed by: _____ Date: _____ File Number: _____

Using the list of potential source categories below, record any indications of contamination or suspected pollutants that may require further investigation or treatment. Sources of contamination may be constant or intermittent or may be linked to single, unrepeated events. For intermittent sources, try to indicate the time of peak activity or contaminant production, including correlations with weather (e.g., wind direction).

Source Category	Checked	Needs Attention	Location	Comments
Special Use/Mixed Use Areas				
Smoking lounges				
Food preparation areas				
Underground or attached parking garages				
Laboratories				
Print shops, art rooms				
Exercise rooms				
Beauty salons				
Redecorating/Repair/Remodeling				
Emissions from new furnishings				
Dust, fibers from demolition				
Odors, volatile compounds				

Pollutant and Source Inventory

Page___of___

Building Name: _____ Address: _____

Completed by: _____ Date: _____ File Number: _____

Using the list of potential source categories below, record any indications of contamination or suspected pollutants that may require further investigation or treatment. Sources of contamination may be constant or intermittent or may be linked to single, unrepeated events. For intermittent sources, try to indicate the time of peak activity or contaminant production, including correlations with weather (e.g., wind direction).

Sections 2, 4 and 6 discuss pollutant sources. Appendix A provides guidance on common measurements.

Source Category	Checked	Needs Attention	Location	Comments

Chemical Inventory

Building Name: _____ File Number: _____

Address: _____

Completed by: _____ Phone: _____

The inventory should include chemicals stored or used in the building for cleaning, maintenance, operations, and pest control. If you have an MSDS (Material Safety Data Sheet) for the chemical, put a check mark in the right-hand column. If not, ask the chemical supplier to provide the MSDS, if one is available.

Sections 2,4 and 6 discuss pollutant sources. Section 4 discusses MSDSs.

Date	Chemical/Brand Name	Use	Storage Location(s)	MSDS on file?